THE MICROTRON

ACCELERATORS AND STORAGE RINGS

A Series of Monographs Edited by John P. Blewett, Brookhaven National Laboratory, and Francis T. Cole, Fermi National Accelerator Laboratory

VOLUME 1
S. P. Kapitza and V. N. Melekhin, **The Microtron**

VOLUME 2
Norman Rostoker and Martin Reiser, **Collective Methods of Acceleration**

THE MICROTRON

S. P. KAPITZA and V. N. MELEKHIN

*Physics Laboratory of the Institute for Physical Problems
Academy of Sciences of the U.S.S.R.*

Translated from the Russian by   I. N. SVIATOSLAVSKY
*Physical Sciences Laboratory
University of Wisconsin, Madison*

English Edition Edited by   EDNOR M. ROWE
*Synchrotron Radiation Center
University of Wisconsin, Madison*

harwood academic publishers
LONDON – CHUR

Copyright © 1978 by OPA, Amsterdam, B.V.

Published under license by:
Harwood Academic Publishers Ltd.
Chansitor House
37/38, Chancery Lane
London WC2A 7EL
   and
Harwood Academic Publishers GmbH
Poststrasse 22
7000 Chur

Editorial Office for the United States of America:
Post Office Box 786
Cooper Station
New York, New York 10003

Originally published in 1969 as МИКРОТРОН by Izdatel'stvo Nauka, Moscow

British Library Cataloguing in Publication Data

Kapitza, S P
 The microtron. − (Accelerators and
 storage rings ; 1).
 1. Cyclotron
 I. Title II. Melekhin, V N
 539.7'37'12   QC787.C8

ISBN 0-906346-01-0

Library of Congress catalog card number 78-67418. ISBN 0-906346-01-0. All rights reserved. No part of this book may be reproduced or utilized in any form or by any means, electronic or mechanical, including photocopying, recording, or by any information storage or retrieval system, without permission in writing from the publishers.

Printed in the United States of America

CONTENTS

Editors' Introduction	*vii*
Translator's Preface	*ix*
Preface	*xi*
Introduction	*xiii*

CHAPTER I - THE PRINCIPLE OF OPERATION OF THE MICROTRON AND TYPES OF MICROTRONS

Section 1 - Introductory Remarks	1
Section 2 - The Principle of Operation of the Microtron	2
Section 3 - Particle Injection	6
Section 4 - Types of Microtrons. Uses of Microtrons	12
Section 5 - The Positron Microtron	20

CHAPTER II - PHASE MOTION IN THE MICROTRON

Section 1 - Small Phase Oscillations	23
Section 2 - Numerical Calculations of Nonlinear Phase Oscillations	32
Section 3 - Conditions of Capture and Resonance Acceleration of Particles	35
Section 4 - The Dependence of Phase Motion on the Thickness of the Accelerating Cavity	39

CHAPTER III - PARTICLE INJECTION CALCULATIONS

Section 1 - Conditions for Injection	44
Section 2 - Equations of Motion	46
Section 3 - Methods of Numerical Calculations	49
Section 4 - Calculations for the First Type of Acceleration	52
Section 5 - Calculations for the Second Type of Acceleration	57
Section 6 - Microtron Efficiency and Means of Increasing It	61

CHAPTER IV - PARTICLE FOCUSING

Section 1 - Vertical and Radial Motion	67
Section 2 - Calculations for Vertical Focusing	72
Section 3 - Calculations for Radial Focusing	78
Section 4 - Analysis of Vertical and Radial Motion	83
Section 5 - Particle Motion in a Slightly Nonuniform Magnetic Field	91

CHAPTER V - CONSTRUCTION OF A MICROTRON

Section 1 - General Remarks ... 99
Section 2 - The Microtron Magnet ... 99
Section 3 - The Microtron Cavity ... 107
Section 4 - Electron Source ... 112
Section 5 - Beam Extraction ... 115
Section 6 - The Accelerator Chamber and the Vacuum System ... 117
Section 7 - High Frequency Generators and Amplifiers and Design of the Waveguide System ... 119
Section 8 - The Microwave System of the Microtron ... 124
Section 9 - Power Supply and Control Systems ... 126

CHAPTER VI - PHYSICAL CHARACTERISTICS OF THE MICROTRON

Section 1 - The Power Needed by the Cavity, and Its Matching to the Waveguide ... 131
Section 2 - Microtron Efficiency ... 134
Section 3 - Choice of the Frequency for the Microtron ... 138
Section 4 - Conditions of Similarity for a Microtron ... 140
Section 5 - Beam-Cavity Interaction ... 141
Section 6 - Coherent Radiation of the Bunches and the Limiting Current in the Microtron ... 147

CHAPTER VII - EXPERIMENTAL STUDY OF THE MICROTRON

Section 1 - Power Measurement in the Cavity and the Current of Accelerated Electrons ... 151
Section 2 - The Structure of the Electron Bunches ... 155
Section 3 - Methods for Analyzing the Focusing ... 160
Section 4 - Results of the Experimental Investigation of Focusing ... 162
Section 5 - The Use of Accelerated Electrons for Obtaining Gamma Radiation and Neutrons ... 175

CHAPTER VIII - CONTINUOUS WAVE MICROTRON

Section 1 - On Transition to a Continuous Wave Mode ... 180
Section 2 - Energy Characteristics of a Continuous Wave Microtron ... 181
Section 3 - Construction of a Continuous Wave Microtron and the Possibilities Thereof ... 184

CONCLUSION ... 187

APPENDIX I - DESIGNING THE MICROTRON ... 190
II - MOTION CHARACTERISTICS AS EXPRESSED BY A SECOND ORDER MATRIX ... 195

EDITORS' INTRODUCTION

The history of particle accelerators and their applications now spans half a century. From very modest beginnings they have become the major tools of nuclear and particle physics. From table top devices they have grown to have lengths and diameters of several kilometers. Their costs have risen to the level of a quarter of a billion dollars.
The past twenty years have seen the development of a first cousin to the particle accelerator--the storage ring, in which a beam of particles can be made to circulate for periods of hours, during which they are made to collide with other charged particles circulating in the opposite direction. Enormous amounts of energy have become available in these "colliding beams" for production of transformation of nuclear particles.
Electrons in storage rings emit strongly in the ultraviolet and X-ray ranges. This "synchrotron light" is finding application in the investigation of the surface and volume phenomena in materials of all sorts.
The history and technology of particle accelerators have not been very well documented. Early discoveries and inventions before World War II are sketchily covered in the physics journals. After the war, beginning in 1956 and continuing to the present, the art has been covered in a series of conferences; some were international, invitation affairs. There have been ten such, alternating between Western Europe, the USSR and the United States. There have been numerous national or local conferences, some covering a wide range of accelerators, others specializing on cyclotrons, "small accelerators" or other more limited topics. The major part of the development of the accelerator art is described in the proceedings of these conferences, having circulation limited to a circle including a few beyond the attendees. One journal has covered the accelerator field since 1970--this is *Particle Accelerators*, an international journal publishing papers on all sorts of accelerators.
It is our hope in this series to provide a more permanent record of the accelerator and storage ring field, more accessible to students, teachers and accelerator specialists. It is with great pleasure that we begin the series with *The Microtron* by S. P. Kapitza and V. N. Melekhin of the Institute for Physical Problems in Moscow.
The microtron is an electron accelerator capable of acceleration of electrons to 50 MeV or higher. It is simple and cheap and very poorly appreciated in countries outside of the USSR. This volume is an updated translation of a book that has made a great impression in the USSR.

Professor Kapitza is well known throughout the world as a major expert on accelerators. He is particularly well known in his home country as a popular exponent of science on television. We are honored to have him and his colleague as our first authors.

John P. Blewett and Francis T. Cole

TRANSLATOR'S PREFACE

With the end of the second World War, there occurred
world-wide a tremendous growth in the expenditure of funds
and effort devoted to research in high energy nuclear physics,
a large part of which was funneled into the development of
the principal instrument of this field: the particle accelerator. Not too surprisingly, considering the highly competitive nature of fundamental research when support of it becomes a part of national policy, the emphasis in the development of particle accelerators was placed on those concepts
most easily extended to higher energy, since higher energy
particle accelerators are both necessary to extend the range
of the measurements and pleasing to the eyes of those charged
with maintaining the national image. Thus the linear accelerator and synchrotron, two accelerator concepts capable of
almost unlimited extension of energy, have emerged as the dominant forms of particle accelerators in the world of high energy physics during the thirty years since the war ended. It
was to be expected then, that a number of elegant and practical low and intermediate energy accelerator concepts proposed during this period would become "lost in the shuffle,"
so to speak, because of a lack of scientific interest and industrial applications in their energy ranges or because of
unsuitability for development into very high energy machines.
Such a machine is the microtron which, when first proposed
by Veksler in 1944, generated a brief flurry of interest among
Western scientists but was then rapidly eclipsed by the electron linear accelerator and the synchrotron as a source for
nuclear and high energy physics research when it became clear
that the microtron as then conceived would probably be limited
to energies of a few tens of MeV.

This was not the case in the Soviet Union, however.
Shortly after the war, for reasons completely unrelated to
its possible application to high energy charged particle research, detailed theoretical studies of the microtron were
carried out independently at two laboratories in Moscow, first
at the Lebedev Institute by S. Kolomenskii and then somewhat
later at the Institute for Physical Problems by S. P. Kapitza
and V. N. Melekhin. Ironically, neither group was to know
of the other's work until the late 1950's.

By 1960, largely due to the work of Kapitza and Melekhin,
the microtron had reached a state of development in the Soviet
Union such that it could equal or surpass the performance of
all other types of electron accelerators at energies up to
approximately 30 MeV. Subsequently, the inherent simplicity,
ease of construction, and low cost of the microtron resulted
in its rapid deployment throughout the Soviet Union for a wide
range of scientific and technological applications which in
the Western nations have traditionally (and expensively) been
fulfilled by betatrons, synchrotrons, and linear accelerators.

During this same period of time, research and development programs on microtrons were carried out elsewhere in the world, most notably in Sweden, Italy, Canada, England, and the United States. However, even though these programs resulted, in some cases, in the construction of successful machines, the development of the microtron did not reach a very high level in the Western world.

In 1970, faced by the certain necessity of finding a suitable and economical replacement for the aged and rapidly deteriorating 50 MeV FFAG synchrotron then in use at the Synchrotron Radiation Center of the University of Wisconsin-Madison, at a time when funding levels for university research programs in the United States were rapidly dwindling, the Radiation Center Operations Group began a serious consideration of the microtron as a possible solution to its requirements. After a period of study, and not without some misgivings, for no successful microtron had every been designed and built in the U.S., the decision was made to proceed with the development of a 35 MeV microtron as the only economically possible course of action. This was because an electron linear accelerator, the only alternative, would cost well in excess of one million dollars.

Upon hearing of our decision to undertake this project, Professor Kapitza was kind enough to send his encouragement and a copy of his and Melekhin's then recently released book *The Microtron*. Upon translation, this book proved to be a remarkable source of information on the history, theory, design philosophy, and operating technique of these machines. And more besides: this book is, in fact, a detailed instruction manual which, if followed carefully, will allow the competent experimentalist with access to the average departmental shop in the average university or technical school to construct an electron accelerator of considerable capabilities of any desired energy up to perhaps 50 MeV. Herein lies the value of this book; it literally makes access to electron accelerators for this energy range available to any investigator determined enough to build one. And this was the reason for the remarkable success of this book in Russia (over 15,000 copies sold) and this was also probably the most important single factor leading to the rapid proliferation of these elegant machines in the Soviet Union.

Unfortunately, our translation of *The Microtron* was not complete in time to save us any great amount of labor in the design and construction of our machines (ultimately we built two: one of 44 MeV and one of 10 MeV energy) but it was reassuring to discover that we had proceeded correctly in our development.

And now, with the kind permission of Professor Kapitza and Dr. Melekhin and through the assistance of Dr. John Blewett of the Brookhaven National Laboratory and the staff of *Particle Accelerators*, especially Ms. Yvonne Winterbottom who typed the final version of the manuscript and gathered together all of the inevitable loose ends, we are able to bring this remarkable treatise to the scientific community.

PREFACE

This book is devoted to the microtron, a cyclic electron accelerator. Work on this accelerator was started at the Physics Laboratory some ten years ago and now the time has come to make a complete report.
Although the microtron was proposed by V. I. Veksler a long time ago, in 1944, initial attempts to build it were not given high priority and as can be seen from this report, fifteen years elapsed before one was built which effectively accelerated electrons. We can now say that in the energy range 5 to 50 MeV, from the standpoint of usefulness, the microtron looks as an effective accelerator. This energy range is also covered by linear accelerators, but the microtron is considerably simpler, is very stable, has a high duty cycle and allows the creation of highly intense electron beams of well grouped bunches.
Initially the microtron attracted our attention by virtue of the fact that the electron bunches it produced appeared very suitable for generating millimeter wavelength radiation. These hopes were not fulfilled, but a result of the effort was the creation of an effective operating microtron, an accelerator which is of interest in many areas of physics and technology. For instance, since it is a simple and effective source of x-rays, it immediately found practical application in industrial radiography and in radioactivation analyses. Being a source of intense electron beams it is used as an injector and as a useful device for a large number of research applications.
Within the last five years the microtron has found wide use abroad as well as in the Soviet Union. Wide acceptance of the microtron as a practical machine came only when, on the basis of new ideas, it was possible to find technical solutions to practical problems which made it into a simple, effective, and dependable device. These ideas are responsible for the rebirth of the microtron. They were mainly obtained at the Physics Laboratory, by the suthors of this report and their collaborators. As will be evident to the readers, the rapid development of the microtron occurred when an effective accelerating cavity was proposed which had a hot electron-emitting cathode. Exact numerical calculations of the electron paths led to the solution of the difficult problem of injection. A theory was worked out which permitted a detailed calculation of the stability of the electron beam.
Growing interest in the microtron brought about a large number of reports devoted to its theory and experimental analysis. We feel that the publication of this book, in which the main results of these efforts are collected and systematized, is timely. This book is written by knowledgeable experts and will undoubtedly raise some corresponding practical

challenges. Some unpublished results obtained by the authors have also been included. Initially, we proposed to publish this book in the series "High Power Electronics." However, inasmuch as this book, to a large extent, is based on formerly published reports, while the series were intended for original works, we decided to release it as a separate monograph.

We hope that this book will be useful for further development of the microtron which still has some unrealized possibilities. For instance, as is mentioned by the authors, it is possible to build a continuous wave microtron of high intensity; the efficiency of the microtron can be increased by the use of superconducting cavities. It is true that one has to be careful not to lose, in the process of such improvements, one of the microtron's main advantages, namely compactness and simplicity.

P. L. Kapitza

Physics Laboratory, Academy of Sciences, USSR

INTRODUCTION

This monograph provides a systematic presentation of the main results in the investigation of the microtron, an electron accelerator of low and intermediate energies.
A microtron is a cyclic accelerator and was proposed (under the name of an electron cyclotron) by V. I. Veksler in 1944, in the very first of his classic papers on relativistic particle accelerators.[41]
The brilliant idea of V. I. Veksler appeared to be timely. In connection with the development of radar, powerful pulsed magnetrons were developed which generated hundreds of kilowatts of power in the centimeter frequency range. Cavities with Q factors of the order of several thousands were proposed. By utilizing the cavity as an accelerating element, excited by the pulsed magnetron, it seemed possible to obtain the necessary energy gain per turn, and thus bring to life Veksler's brainchild.
This was first accomplished by a Canadian group who in 1948 constructed the first accelerator of this type for an energy of 4.8 MeV[42,43] and called it the microtron. This name is derived not from the small size of the device, but rather from the fact that the centimeter wavelength bands in English and American literature are referred to as *micro*waves. In subsequent years several other microtrons were built, and some theoretical works dealing with more thorough analysis of particle motion in the microtron were published.
However, these accelerators did not receive wide acceptance because of the low electron intensities produced in them, amounting to 1-2 mA, in pulses ∿ 1 μsec. At the same time, the electron linear accelerator was rapidly developed, driven by powerful microwave generators. In some respects, the design of the linear accelerator is more complex than that of the microtron, but the acclerated electron current amounted to tens or hundreds of milliamperes, i.e., several orders of magnitude greater than that in the microtrons. This led to greater interest in linear accelerators and the microtron was considered to be less promising.
In the Physics Laboratory of the Academy of Sciences, USSR (Institute for Physical Problems) work on microtrons was started more than ten years ago in connection with a general interest in relativistic methods for generating very short radio waves, stimulated by the problem of developing millimeter and submillimeter electronics. One of the ideas of relativistic electronics lies in the use of coherent radiation from electrons when they are grouped together during acceleration to high energies into compact bunches. At that time a number of laboratories were experimenting with relativistic electrons in linear accelerators. We, however, decided to say with the microtron in spite of the lower currents

xiii

achieved in it at that time, as compared to the linear accelerator, attracted by the idea that in principle, the microtron has a high degree of phase stability (as compared to the linear accelerator). It is true that the area of phase stability in the microtron is smaller and phase movement is tighter than in the linac; besides, the microtron accelerating cavity exerts strong focusing forces on the electrons in the vertical and radial directions. Thus, it was reasonable to expect that the microtron will produce more compact electron bunches which are indispensable for all conceivable experiments in relativistic electronics. At that time it was pointed out (by one of the authors of this book) that it would be possible to bunch the particles while they moved in the magnet field and simultaneously accelerate them in a cavity similar to a microtron.[1]

Taking these considerations into account we started to build and investigate the microtron, in spite of the seemingly discouraging results of the formerly built machines of this type. The effort expended since that time has allowed us to build an effective microtron. This became possible only after solutions to a series of problems were obtained.

New methods for injecting particles into the microtron were proposed and investigated, and a theory of focusing electrons in the microtron was developed. Of great significance was the use of numerical methods for calculating the movement of particles in the microtron. Numerical methods were essential for the microtron as one cannot consider the fields to be small as in other machines. The number of orbits in the microtron is small but the energy gain per revolution is large; for this reason calculations based on the perturbation theory are not always applicable. If we want to obtain results which to some degree of accuracy describe the capture and movement of particles in the microtron then we have to resort to numerical analysis.

Just as important was the solution of many experimental and technical problems. One of the most significant developments was the use of effective thermocathodes made of lanthanum boride and the introduction of accelerating cavities which develop strong electric fields of the order of 1 million V/cm. For exciting the accelerating cavities, we used powerful pulsed magnetrons which were initially developed for radar; although they turned out to be useful for accelerators, there was the problem of matching them to the cavity. Finally, magnets were developed which produced fields of the necessary shape and uniformity.

The microtron which we constructed was the first one to be built in the Soviet Union; it started to work in 1958.[2,3] This microtron had 12 orbits with a chamber of 70 cm in diameter; electrons were initially accelerated to 6 or 7 MeV and later to 12 MeV. The machine was used for developing and testing new ideas which were later incorporated in other, larger, microtrons.

In 1961, a large microtron of 30 orbits was built at the Institute for Physical Problems which had a chamber 110 cm in diameter and was to have an energy of 30 MeV.[10,19] A similar machine was also built simultaneously in the laboratory of

Neutron Physics of the Joint Institute for Nuclear Research at Dubna,[28] as a joint project with the Institute for Physical Problems. The combination of the Dubna microtron with a pulsed reactor for fast neutrons led to the operation of one of the most powerful neutron sources in our country.

The first microtron has subsequently been disassembled and for a period of several years was exhibited at a national industrial exhibition. In its place was built another microtron which incorporated several significant improvements, resulting in a simple and reliable accelerator of 17 orbits, which accelerated electrons to energies of 10 to 11 MeV.[27]

At present this machine is being used for nuclear physics research,[34-36] of which one of the most interesting is the study of angular asymmetry in photofission of heavy nuclei. These investigations have led to the discovery of the quadrupole channels of fission in even-even nuclei and of the anomaly of mass asymmetry at low excitation energies. These results were made possible only by the high intensity and high energy resolution of the electron beam. They have also demonstrated the importance of having intense sources of monochromatic electrons for the study of threshold events in nuclear physics.

On this microtron experiments were also conducted in relativistic electronics, when the radiation produced by the passage of electron bunches through open cavities was studied.

At this time we must say that the hopes we had for generating very short waves by means of relativistic electrons have thus far not been realized. On the one hand, quantum generators such as masers and lasers were invented and developed. On the other, it became clear that effective radiation by fast electrons and their bunches formed by the accelerator is not very likely to be possible in the absence of feedback from the radiated field to the electrons; such feedback would bring about phase focusing. For high energy electrons, phase focusing can exist only with high intensity radiation which, thus far, has not been obtained. Lower power, which is of interest primarily in radiospectroscopy and communications can be obtained by ordinary (nonrelativistic) high frequency electronics. At the present time, backward traveling-wave tubes readily generate waves down to $\lambda = 0.3$ mm. The orotron, invented in our laboratory by F. S. Rusin and G. D. Bogomolov, successfully combines in itself the open cavity with the principles of classic electronics and promises to generate even shorter waves.[32,39]

As a result of these circumstances, interest in relativistic electronics diminished. Nevertheless, we should keep in mind the possibilities which relativistic electronics have, especially for generating very high power. These possibilities have not been fully investigated and their potential is not completely realized. Actually, the microtron itself is an arrangement of relativistic electronics in which, however, high frequency oscillations are used for accelerating particles but are not generated, amplified, or transformed into oscillations at other frequencies.

In our work, we were able to develop the theory of the accelerator and to build it. In contrast with other contemporary accelerators whose sheer size precludes the possibility of doing detailed experimentation, the microtron made it possible not only to perform the calculations, but to try out experimentally the various methods of construction and investigate the different modes of operation, etc. After we constructed in our laboratory an effective microtron with high beam power, several other institutes in our country also built accelerators of this type, primarily for solving various physics and technical problems.*

Because of the special quality of its beam, the microtron has found use as an injector into synchrotrons. It was first used as an injector into the synchrotron at Lund (Sweden). In the USSR, the microtron was used as an injector at the Lebedev Physics Institute of the Academy of Sciences, and at Tomsk at the synchrotron of the Tomsk Polytechnical Institute. Work on the microtron at the Lebedev Physics Institute produced an effective method of accelerating positrons and their injection into a storage ring. Recently, a microtron injector of 7 to 10 MeV is being built for the synchrotron of 1 to 2 GeV at Frascati (Italy).[93]†

It is now clear that the microtron is no longer a rarity which exists for merely demonstrating the principle of its operation, but has come of age as an effective particle accelerator. It has become an efficient, useful and dependable accelerator of electrons for low and medium energies. The physics and technical principles of its construction are clear. These circumstances have prompted the authors to write this book in which results obtained over the past years are systematized.

This book is based on research performed at the Physics Laboratory of the Academy of Sciences, USSR (Institute for Physical Problems). Former results on the microtron (work published before 1960) can be found in a review by A. P. Greenberg.[44] We refer to these reports only to the extent that they will be pertinent to the problems that confront us.

Many of the theoretical results in this book are presented without great detail and it is suggested that the reader refer to the original work. Nevertheless, all the original results obtained during these years, including the results which were not published earlier and contained only in internal laboratory reports are included here. In the list of references found at the end of this book numbers 1 to 40 cover work published at the Physics Laboratory that are connected with this work.

In Chapter I we describe the principle of operation of the microtron and give its general description. Subsequent chapters are devoted to particle dynamics of the microtron. The theory of phase oscillations in the microtron, numerical methods for calculating the injection and consequent movement of the particles, and the theory of vertical and radial

---

*More than 36 microtrons are now successfully operating in the Soviet Union. [Note added 1976.]

†A description of the first American modern microtron built at the University of Wisconsin can be found in Ref. 100.

focusing as well as the influence of nonuniformity of the
magnetic fields on the motion of the electrons are presented.
   In Chapter V the microtrons built at the Physics Laboratory are described; this chapter draws mainly on our own experience. Considerations on the choice of the high frequency
cystem of the microtron are given.
   The next two chapters cover the energy characteristics
of the beam and an evaluation of high current effects as well
as results of experimental possibilities in nuclear physics
and technology.
   In the Appendices an illustration of the engineering design for a microtron are provided as well as some mathematical
derivations concerning the theory of focusing.
   It must be mentioned here that Chapters I, V, VI, and
VIII, Appendix I, and Sections 2 and 5 of Chapter VII were
written by S. P. Kapitza, while Chapters II, III, IV, Appendix II, and Sections 1, 3, and 4 of Chapter VII were written
by V. N. Melekhin.
   This book summarizes the development of the microtron.
However, the development of this accelerator will not stop
at this stage even though we can say that the main problems
have been solved for small machines in the energy range from
10 to 30 MeV. The number of such machines will depend
strongly on their commercial value as well as the actual need
for them.
   On the other hand, undoubtedly unique, large microtrons
operating in a cw mode or in the regine of long pulses will
be built. Such machines are valuable as an instruments of
nuclear research since they cover an important energy interval from several million electron volts to ~ 100 MeV.
   The main problem which the authors placed before
themselves when they wrote this book, was the assimilation
and discussion of the results so far obtained. Without in-
sisting on our own prognosis the authors only hope that this
attainment will be a step forward. Only the future will show
whether the microtron will withstand the strong competition
with other machines and find its own niche in the world of
accelerators.
   As a chapter in accelerator physics the microtron will
undoubtedly remain. Will this book be read? This depends
on the degree to which microtrons will be built, which in
turn depends on the readers of this book.

CHAPTER I

THE PRINCIPLE OF OPERATION OF THE MICROTRON AND TYPES OF MICROTRONS

SECTION 1 - INTRODUCTORY REMARKS

In this chapter we will present the principle of operation of the microtron and, in general, characterize the main types of microtrons.

The microtron in many ways resembles the common, Lawrence, cyclotron. As in this first cyclic accelerator, particles in a microtron move in a constant and uniform magnetic field. The particles are accelerated by alternating electric field of constant frequency. For this reason, as was previously mentioned, V. I. Veksler on proposing this type of accelerator, called it the electron cyclotron.[41] However, in contrast to the cyclotron, the microtron was proposed as an accelerator of relativistic particles, even of ultrarelativistic particles, the energies of which are essentially greater than the rest energy. For this reason, in its present form, the microtron is only useful for accelerating light particles - electrons and positrons.

The microtron is the first cyclic relativistic accelerator in the dynamics of which the principle of phase stability was discovered.

> This automatic phasing - present by virtue of the fact that the time interval between two successive accelerations depends on the accelerating voltage, is a general characteristic of accelerators of this type allowing (at least in principle) the acceleration of particles by many different methods, even in cases when the magnetic field rises with time.[41]

Thus, did Veksler conclude his first paper[41] on the new method for accelerating relativistic particles.

We will not investigate further the development of the principle of phase stability, which brought about the development of synchrotrons, synchrocyclotrons, and other accelerators, but will limit ourselves to the questions concerned with the electron microtron. The development of the microtron was somewhat slow in comparison with other accelerators. Nevertheless, this comparatively small machine, effectively and simply accelerates electrons to energies of the order of 10 MeV and is very interesting both from the practical and the theoretical standpoint. Actually, a detailed theoretical study of its operation is instructive for the general theory of cyclic accelerators and provides for interesting development in nonlinear mechanics.

SECTION 2 - THE PRINCIPLE OF OPERATION OF THE MICROTRON

In the microtron, particles are accelerated by an alternating electric field of constant frequency in a constant uniform magnetic field. We will mainly concern ourselves with the movement of electrons, while specific problems concerned with the acceleration of positrons will be treated in Section 5 of this chapter.

In the vacuum chamber electrons follow circular paths with a common tangent point (Fig. 1.1). At this point the

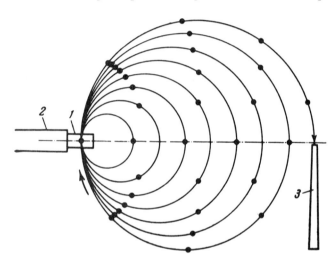

Fig. 1.1 Particle motion in a microtron. (1) Accelerating cavity; (2) Waveguide; (3) Magnetic channel for particle extraction.

cavity which supplies the high frequency electric field needed to accelerate the electrons is placed. The cavity is excited by a powerful source of high frequency power, usually a pulsed magnetron with a power of at least hundreds of kilowatts.

After each passage through the cavity, the electrons gain a certain amount of energy ΔU and pass on to the next orbit. When the electrons reach the last orbit, they either hit the internal target or else pass through a magnetic channel and are extracted.

Synchronism of electron motion with the accelerating field is achieved by the fact that each succeeding orbit is longer than the former one by an integer number, q, periods of the high frequency oscillations. The ratio of the orbit period of the particle to the period of the accelerating field is commonly known as the harmonic number of the

MICROTRON

accelerating mode. Since the length of each orbit is different, the microtron is an accelerator which has a changing harmonic number.

It is common to have accelerating regimes where g = 1; we will consider it first.

The time of revolution T of a relativistic particle in a magnetic field H is determined by

$$T = \frac{2\pi}{ec} \frac{U}{H} ,\qquad(1.1)$$

where U is the total energy of the particle. Thus, the condition of synchronism has the form

$$\Delta T = \frac{2\pi}{ec} \frac{\Delta U}{H} = T_o ,\qquad(1.2)$$

where ΔU is the particle energy gain during each passage through the cavity, ΔT is the corresponding change in time on the new orbit, and T_o is the period of the high frequency field in the cavity. When the conditions of (1.2) are satisfied the electrons pass through the cavity during the same phase of the accelerating field (see Fig. 1.2). It is con-

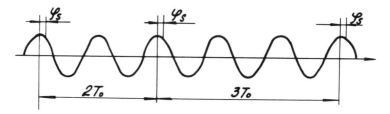

Fig. 1.2 The change in the harmonic number in a microtron.

venient to change (1.2) into the form

$$\Delta\Gamma = \frac{H}{H_o} ,\qquad(1.3)$$

where we have introduced $U_o = mc^2$, the rest energy of the particle, and through Γ we have designated its total energy measured in units of the rest energy ($\Gamma = U/U_o$, $\Delta\Gamma = \Delta U/U_o$); for an electron $U_o = 0.511$ MeV. The magnetic field in formula (1.3) is

$$H_o = \frac{2\pi U_o}{ecT_o} = \frac{2\pi mc^2}{e\lambda}\qquad(1.4)$$

is the cyclotron field; it is determined by the period of the accelerating field (or by the wavelength $\lambda = cT_o$, or by the frequency $f = 1/T_o$). The cyclotron frequency is a measure of the magnetic field H and thus it is reasonable to introduce the parameter $\Omega = H/H_o$.

Microtrons, like most conventional linear accelerators, operate in the 10-cm waveband. At a frequency of 3000 MHz (λ = 10 cm) the cyclotron field H_o = 1070 Oe.

In this way the condition of synchronism (1.3) can be given the following form:

$$\Delta\Gamma = \Omega \qquad (1.5)$$

Thus the parameter Ω appears to be the main characteristic of the accelerating mode and we will see that many (if not all) properties of the accelerator are connected with this parameter.

In the microtron usually $\Delta\Gamma \sim 1$, and it is thus that the microtron differs from all other cyclic accelerators. In all other accelerators, one can consider the energy gain per revolution as compared with the energy expended during the time of acceleration to be small. Thus, the number of revolutions in them is very great, while the number of orbits in the microtron is considerably less; in the presently constructed machines, it does not exceed 60.

With the conditions $\Delta\Gamma \sim 1$, λ = 10 cm, and effective length of the accelerating gap of 1.6 cm, we have E = 320 kV/cm; in the absolute system of units E = H = H_o. For greater values of Ω the electric field is correspondingly larger, inasmuch as E, H and ΔU are proportional to Ω.

Thus, with each passage through the accelerating cavity the electron energy must increase by a value about the same as the rest energy. The creation of such an accelerating field is the main problem of building a microtron; this problem was solved with the aid of powerful generators in the centimeter waveband and cavities with Q of the order of 10^4.

The acceleration of protons and heavier charged particles in conventional microtrons is impossible since it requires a high frequency field thousands of times more intense. Several possibilities for operating at high harmonic number for accelerating protons were investigated by Roberts.[4,5]

The total energy of the electrons at the nth orbit is equal to

$$U_n = (n - 1)\, \Omega U_o + U_1 , \qquad (1.6)$$

where U_1 is the energy at the first orbit completely encircling the cavity. Inasmuch as the time of flight at the nth orbit must take an integral number of periods T_o, let us say $(n + m - 1)T_o$, then the energy U_n must be equal to

$$U_n = (n + m - 1)\, \Omega U_o . \qquad (1.7)$$

From (1.6) and (1.7) we can determine the total energy at the first orbit

$$U_1 = m\Omega U_o \qquad (1.8)$$

and the corresponding kinetic energy

$$U_1 - U_o = (m\Omega - 1)\, U_o . \qquad (1.9)$$

MICROTRON

The energy at the first orbit depends only on the parameter Ω and the integer m which determines the length and the dimension of the first orbit encircling the cavity. Since $U_1 > U_0$ not all values m, n and Ω are possible and the smallest value of Ω is

$$\Omega_{min} = \frac{1}{m} . \qquad (1.10)$$

Since the motion of the particles is ultrarelativistic ($\beta \approx 1$) the diameter of the nth orbit is

$$D_n = (n + m - 1) \frac{\lambda}{\pi} \qquad (1.11)$$

(we designate n = 1 for the first orbit completely encircling the resonator). On the following orbit the diameter increases by the factor

$$\Delta D = \lambda/\pi . \qquad (1.12)$$

If the integer g is arbitrary then the above-written equations take on the following form

$$\Delta \Gamma = g\Omega , \qquad (1.5')$$

$$D_n = [m + g(n - 1)] \frac{\lambda}{\pi} , \qquad (1.11')$$

$$\Delta D = g \frac{\lambda}{\pi} , \qquad (1.12')$$

while the other equations remain unchanged. The derivations of these expressions is elementary.

Due to the principle of phase stability, stable motion is possible not only for particles which satisfy exactly these conditions but also for particles for which these expressions are true only on the average.

Phase motion in the microtron was first investigated in 1950 in a Thesis by A. A. Kolomensky.[4,6] This and later reports have shown that the region of phase stability is not large, for instance when g = 1 the equilibrium phase ϕ_s (see Fig. 1.2) can lie within the boundary 0 to 32°. For g = 2, 3, ... the region of phase stability is maller (approximately by a factor g). For this reason operating modes where g = 2, 3, ... are avoided, since even when g = 1 stable phase motion exists only in a small interval of the initial phase.

Phase oscillations are periodic functions of the orbit number and their period usually covers 4-5 orbits. They are not periodic in time and their duration grows as the size of the orbit increases. It is interesting to note that the phase oscillations as well as the orbital motion are periodic in time in the reference system of the electron. In other words, from the point of view of the observer moving with the electron the frequency of the accelerating field increases linearly in time.

It is possible to introduce more complex accelerating modes where the average phase of the electron bunch changes from orbit to orbit.[62] These modes are also phase stable, but the region of phase stability seems to be quite small (of the order of 1°). Such modes can be considered as alternating phase focusing: the net result of a series of phase movements, some of which are unstable, together give stable phase motion. These complex modes of operation have not yet found any application.

The constant magnetic field in the microtron is uniform and particles advancing in it always return to the cavity. Vertical focusing* in the microtron is exclusively determined by the electromagnetic field in the cavity. In contrast to the linear accelerator the fields in the cavity apertures produce the focusing action. This follows from the fact that the focusing forces at the entrance hole are stronger than at the exit as long as the stable phase region lies in the first quadrant (see Fig. 1.2). The focusing force depends on the shape of the aperture in the cavity (see Chapter IV).

Radial oscillations in the plane of the orbit are determined by the combined action of the constant magnetic field and the high frequency cavity field. It is interesting to note that one can make the magnetic field slightly nonuniform so that it falls off on moving away from the common orbit diameter; such a gable-shaped field exerts more effective radial focusing.[23]

In the microtron the amplitude of the phase oscillations and the size of the beam are determined by the stable-phase region and by the vertical aperture. The limits of the stability region are small and thus lead to electron bunches 5-8 mm long, a height of 1-5 mm, and a radial width of 3-6 mm for a $\lambda = 10$ cm. The area in which stable particle motion can exist is small and thus the electron beam in the microtron is well collimated and highly monoenergetic. However, the small size of the phase space of the accelerated particles makes it difficult to inject the particles. This difficulty for a long time impeded the development of the microtron, since it led to beams of low intensity.

SECTION 3 - PARTICLE INJECTION

In the first (Canadian) microtron[42,43] a toroidal reentrant accelerating cavity similar to the one proposed by V. I. Veksler was used. This cavity was excited by a pulsed S-band magnetron with a pulsed power output of 300 kW. The accelerating gap was 0.08-0.10 λ (approximately 9 mm). The electric field in this gap near the sharp corners of the cavity was of the order 10^6 V/cm. Under the action of this high frequency field electron emission took place directly from

---

*Following the traditional accelerator convention we designate the vertical direction as being perpendicular to the median plane of the accelerator which is usually horizontal. The latter, however, is not necessarily true: for instance, in the microtron of the Joint Institute for Nuclear Research, Dubna, the median plane is vertical!

MICROTRON

from the edges of the copper walls of the cavity (Fig. 1.5). Although the emitted current was large, reaching 1 A, only a small part of the electrons were captured in the accelerating mode and the beam current did not exceed 1 mA.*

For injection from the cavity $U_1 = U_o + \Delta U$ such that when $g = 1$ and $m = 2$ the energy at the nth orbit was equal to $U_m = (n + 1) U_o$, i.e., a multiple of the rest energy. In this case $\Omega = 1$. In the first microtrons an attempt at acceleration was made with $m = 3$ and $\Omega = 1/2$ in the so-called 1/2 mode with lower values of radio-frequency power and magnetic field.

These modes of acceleration were studied in detail by the London group which constructed and reported on a microtron with 12 orbits and an energy of 6 MeV.[49] However, the current in the beam in this machine was also small: 1-2 mA with a pulse length of 1.5 μsec. Following the example of the Canadian and the London microtrons there were several other machines built,[51-53] however, substantially larger currents were not obtained.

Kaiser[50] built several microtrons operating in the 3 cm wavelength; these microtrons also utilized toroidal cavities. These accelerators operated in the half mode; their currents were low (of the order of 0.1 mA) and these efforts did not lead to any further development.

The reports mentioned thus far have shown, in principle, the possibility of accelerating electrons in a microtron, by demonstrating their stable motion. However, these reports do not give an adequate solution to the problem of injection. The absence of an effective injection system impeded the increase of the current in the beam and consequently the development and wide introduction of microtrons. To increase the current several injection methods were proposed. Detailed reports of these methods are contained in a review by A. P. Greenberg[44] we will briefly consider them pointing out their merits and deficiencies.

The most obvious method is the use of a hot cathode situated close to one of the edges of the accelerating gap. A hot cathode was experimentally tried out,[53] however, the authors were not able to increase the accelerated current appreciably.

Paulin[54] proposed an injection method schematically presented in Fig 1.3. It is based on the use of a toroidal cavity; the radius of the first orbit is small and thus, passes through the center of the cavity in the wall of which is an additional gap for the electrons to exit from. The flight time on this internal orbit is equal to one period of the high frequency field and thus, the electrons returning to the cavity are in the same phase. The author supposes that this method can be used with an acceleration mode $\Omega = 4$. However, no calculations or experimental attempts verifying this

---

*Here and in the future when we mention current, we have in mind the pulsed value. The duty ratio of the majority of microtrons (including the abovementioned ones) is close to 1000 such that for a pulsed current value of 1 mA, the corresponding average current in the beam is approximately 1 μA.

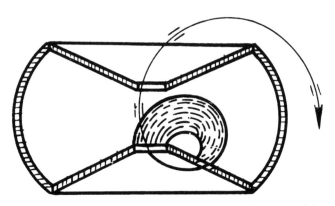

Fig. 1.3   Injection method proposed by Paulin.

conclusion are presented.

Schmeltzer[55] proposed the use of two cavities, one serving as the injector into the other. Aitkin[60] proposed a cylindrical cavity in the center of which a source of electrons is situated, and on the first turn around the cavity the particles are shielded inside magnetic channels. It was suggested that by this method it would be possible to accomplish acceleration in a magnetic field which was double in strength.

The new injection method was proposed and accomplished in practice by O. Wernholm[59] who built a 1.2-GeV strong focusing synchrotron which used a microtron injector. In this microtron electrons were introduced into a toroidal cavity by a pulsed high voltage coaxial gun (Fig. 1.4). On the tenth orbit in the microtron at an energy of 5 MeV beam current of 60-70 mA was obtained. This injection method increased the accelerated current by an order of magnitude, however, the magnetic field in the accelerator as before was equal to the cyclotron field and the energy of the accelerated particles could not be changed smoothly.

In the Physics Laboratory of the Academy of Sciences, USSR, initial experiments were made on a conventional microtron with a toroidal cavity and electron emission under the action of the accelerating field. Electron trajectories in this cavity obtained by G. P. Prudkovsky on his trajectograph[31] are shown in Fig. 1.5. In these experiments, conducted by V. P. Bikov, cavities with gaps of various shapes were used and the dependence of the emitted current on the condition of the cavity surfaces was studied. By purely empirical methods it was possible to increase the current (to 7 mA on the 12th orbit), however, the results were hard to reproduce. This forced us to seek new methods of injection.

In 1958, an attempt to use double cavities was made (Fig. 1.6). The supplementary cavity was joined to the main cavity by a coupling loop. In such an arrangement, it is difficult to insure the necessary amplitude and phase relationships between the fields in the cavities and, on the whole, this system turned out to be practically inoperable.

MICROTRON

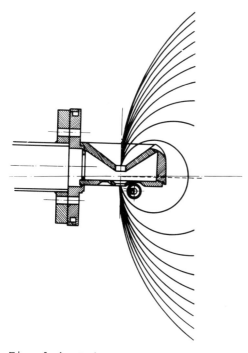

Fig. 1.4 Injection method proposed and carried out by Wernholm.

Two new methods of injection were proposed in 1959 by V. N. Melekhin; the corresponding particle trajectories are shown in Fig. 1.7. For both types of acceleration the cavities used had a thickness which was several times greater, and the alternating electric field away from the center dropped off less rapidly, than in the re-entrant cavities used previously. This idea allows the high frequency field to accelerate the electrons inside the cavity during the first half revolution, which is essential for injection.

Cavities for acceleration of the first or second type normally take the form of a circular cylinder the diameter of which is several times larger than the height.[2,3] Mode E_{010} is used and such a cavity is similar to a single section of a linac. Electron emission in this cavity takes place from a hot cathode placed on one of its walls.

The frequency of the mode E_{010} depends only on the diameter of the cavity $2a$. $2a = 0.735 \lambda$, thus the resonant length of the wave does not depend on the thickness of the cavity; normally, this thickness is $\lambda/5$ to $\lambda/4$. Such cavities we call flat cavities.

In the first type of acceleration the emitter is placed approximately in the middle of the cavity radius. In the second type, electrons start close to the axis of the cavity and, following a more complex trajectory, leave it from a

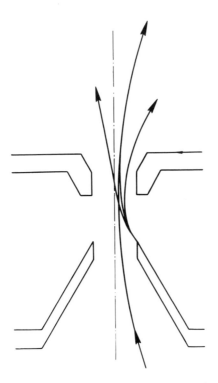

Fig. 1.5 Electron trajectories in a toroidal cavity (obtained on G. P. Prudkovsky trajectograph).

Fig. 1.6 Dual cavity.

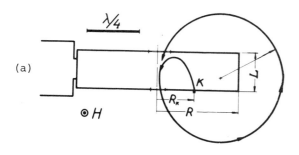

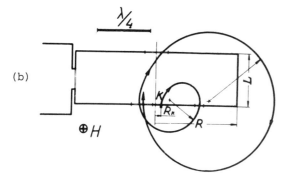

Fig. 1.7 Injection methods proposed by V. N. Melekhin: (a) First type of acceleration; (b) Second type of acceleration.

supplementary hole (Fig. 1.7).

For the first type of acceleration, m = 2, and consequently the first revolution takes 2 periods of the accelerating field; thus, the total electron energy after the first orbit is equal to $U_1 = 2\Omega U_0$. In the second type of acceleration, the electron leaving the supplementary hole must have an energy of ΩU_0, and only after passing through the cavity for the second time, does its energy increase to $2\Omega U_0$. Calculations and experiments have shown that particles can be accelerated over a wide range of Ω, thus for the first type with m = 2, $0.65 < \Omega < 1.7$, while for the second type $1.7 < \Omega < 2.2$. We must point out that with these methods of injection one can smoothly change U by changing Ω within the same accelerating cavity and the same position of the emitter. This permits one to vary the energy of the beam over a large range by merely changing the magnetic field and the high frequency electric field in the cavity. One advantage of flat cavities appears to be a smaller electric field as compared to the toroidal cavities since the gap in them is larger, being between 18 and 25 mm for λ = 10 cm.

Due to their simplicity of form, field distribution in flat cylindrical cavities is well known. Thus, electron motion can be calculated by numerical integration of the equation of motion. Methods and results of these calculations performed on computers are presented in Chapter III. In this way, one can calculate many modes of acceleration for cavities of various configurations, not only circular cylindrical, but also rectangular, H-shaped as well as cavities bounded by spherical surfaces.

Experiments on the modes of acceleration have fully confirmed the results of the corresponding calculations and the design of new modes of acceleration and of new cavities is based on numerical calculations performed on computers.

SECTION 4 - TYPES OF MICROTRONS

Uses of Microtrons

The majority of contemporary high-current microtrons utilize flat cavities and the types of acceleration (the first and the second) investigated above. The current of accelerated particles reaches 100 mA or more.

Most of the microtrons thus far built operate in the 10-cm waveband (S-band). In Table 1.1 we present the parameters of some of the more important microtrons; one must differentiate accelerators of relatively low energy (from 10 to 15 MeV, see Nos. 1, 4, 5, 6) from accelerators of high energy (up to 30 MeV, see Nos. 2, 3, and 7).

In Fig. 1.8 is given a general view of the first microtron built at the Physics Laboratory of the Academy of Sciences USSR (Institute for Physical Problems); a schematic of it and dimensions are given in Fig. 1.9. It was later disassembled and in its place a 17 orbit microtron was built. This accelerator is excited by a standard pulsed magnetron with an

Fig. 1.8 General view of the first microtron (Institute for Physical Problems).

TABLE 1.1 Microtron in the 10-cm waveband.

	Institute	Pole Diameter (cm)	Particle Energy (MeV)	Number of Orbits	Pulse Current (mA)	Parameter (Ω)	Weight of Magnet (tons)	Use
1.	Institute for Physical Problems Moscow, USSR (Ref. 27)	75	15	17	35	1-2	0.9	Nuclear physics and relativistic electronics
2.	Institute for Physical Problems Moscow, USSR (Ref. 19)	110	32	30	50	1-2.2	5	Particle dynamics, radioactivation analysis
3.	United Institute for Nuclear Research, Dubna, USSR (Ref. 28)	110	30	30	60	2	4.5	Injector into a pulsed reactor for fast neutrons
4.	Physical Institute Ac. Science USSR, Moscow, USSR (Ref. 66)	60	7	10	110	1.2	2	Synchrotron injector, positron acceleration
5.	Lunde University, Sweden (Ref. 59)	50	6.4	10	50	1.05	0.6	Synchrotron injector
6.	Western Ontario University, Canada (Ref. 86)	50	6.2	8	40	1-3	Split	Particle dynamics, relativistic
7.	London University, England (Ref. 60)	200	29	56	10^{-3}	1	20	Work discontinued
8.	University of Wisconsin P.S.L., USA (Ref. 100)	137	44	34	15	2.5	15	Storage ring injector

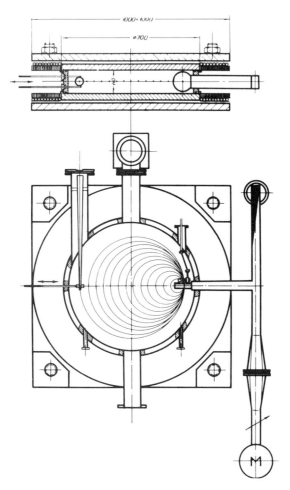

Fig. 1.9 Schematic of the first microtron.

average power output of 2.5 kW that produces a 15-MeV beam. The average beam power is 0.5 kW.

This accelerator (the general design is shown in Figs. 1.10 and 1.11 and which is described in detail in Chapter V) is at present being used for scientific and technical purposes. Its operation is simple and reliable, the beam is well focused and the vertical and radial oscillations give a focus spot 2×3 mm^2. The intensity of bremsstrahlung radiation at a distance of 1 m from the tungsten target 1 mm thick is approximately 2000 R/min at a beam energy of 10 MeV. The small dimensions of the beam and the high γ-radiation intensity make this a high intensity accelerator very suitable for industrial radiography.

Fig. 1.10 General view of the 17-orbit microtron.

Large microtrons up to 30 MeV constructed at the Institute for Physical Problems and the Laboratory of Neutron Physics at Dubna have much in common as far as construction and have similar parameters (see Refs. 10, 19, and 28). Both accelerators have 5-ton magnets of armor type. The microtron of the Institute for Physical Problems (Fig. 1.12) has been extensively used for beam dynamics studies and as a source of bremsstrahlung radiation for activation analysis. The Dubna microtron has been used as an injector to a pulsed reactor for fast neutrons (IBR). Operating in conjunction with the microtron, the IBR becomes a dynamic subcritical system with very high multiplication efficiency as the neutron pulse obtained from the microtron is multiplied by 150 or 300 times in the uranium-plutonium subcritical assembly of the reactor.[28] However, in spite of the effectiveness of this combination it cannot be considered that the microtron is best suited for generating neutrons.* It seems that for this particular application, powerful linear accelerators transiently operating by storing high frequency energy are more suitable. Nonstationary operation leading to a substantial increase in the current while shortening the pulse is difficult to accomplish in the microtron since the amplitude of the high frequency field in it, according to Eq. (1.5), is strictly determined by the magnetic field. The wide energy spread of the linear accelerator is inconsequential for the production of neutrons.

Of all the microtrons known to us, the one with the largest number of orbits is the 2-m London microtron constructed in 1956.[60] It had 56 orbits, however the current in the beam was very low (0.05 mA). In this accelerator electrons were

---

*At present the microtron at Dubna has been dismantled and a 200-mA, 35-MeV linear accelerator installed (1972).

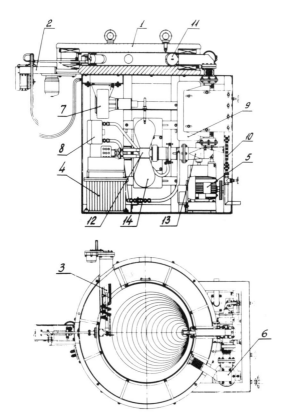

Fig. 1.11 Schematic of the 17-orbit microtron: (1) Magnet pole; (2) Extraction mechanism; (3) Extraction tube; (4) Pulsed transformer; (5) Forevacuum pump; (6) Diffusion pump; (7) Fan; (8) Air-cooling; (9) Ferrite isolator magnet; (10) Calibrator; (11) Cavity; (12) Magnetron; (13) Waveguide; (14) Magnetron magnet; (15) Cavity tuning mechanism.

emitted by the walls of a toroidal cavity and in the process of acceleration large numbers of particles were lost due to the lack of effective focusing. Of particular interest in this accelerator was the magnet, in which, thanks to a very thorough design, a high field uniformity was achieved. For correcting the location of the orbit, coils were placed on the poles, without which, in general, it would have been impossible to get the beam to 56 orbits. At present, as far as we know, the work on this accelerator has been discontinued.

All of these accelerators operate in the 10-cm waveband, in the pulsed mode, with pulse lengths of 1 to 3 μsec. The duty ratio of the high frequency generators was usually equal to 1000 and most microtrons operate with this duty ratio. A larger duty ratio is usually assigned only to injectors in

MICROTRON   17

Fig. 1.12   General view of the 30-orbit microtron.

which the pulse rate (12.5 to 50 Hz) is determined by the main accelerator.

In microtrons the magnetic field is not only constant in time but is uniform in space. However, microtrons can also use azimuthally nonuniform magnetic systems resulting in the so-called split microtrons. Such systems were proposed by Moroz[62] and independently by Roberts.[45] Such a split system is shown in Fig. 1.13: the shaded areas have a uniform magnetic field while the remaining areas have zero field.

The particle trajectory consists of straight and curved sections. The cavity is placed in a plane of symmetry within

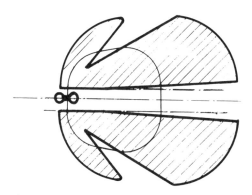

Fig. 1.13   Schematic of the racetrack microtron with edge focusing.

the magnet, in a region where there is no magnetic field. For motion in a field such as this it is possible not only to have microtron mode of acceleration, but also (due to edge effects) to insure particle focusing in the radial and vertical directions.

A small split microtron with 7 orbits and an energy of 5 MeV was built in Canada (see Refs. 63 and 86 and Fig. 1.14).

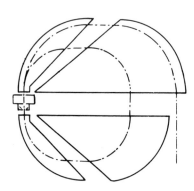

Fig. 1.14 Schematic of the Canadian racetrack microtron.

This accelerator operated in the 10-cm waveband; it utilizes a flat cavity while the electrons with an initial energy of 20 keV are injected by an external gun. The current at the 7th orbit reached 40 mA. The magnetic field in this accelerator needed to be very carefully corrected by adjusting the position of the pole pieces. In the central part of the magnet in order to reduce stray fields it was also necessary to use additional shielding.

This work has demonstrated in principle the feasibility of the split microtron (a similar accelerator has since been constructed at Berkeley[64]) but has shown that the creation of the field itself is quite difficult. In machines with uniform fields, which are easy to build and control, stability of motion can be insured by means of apertures in the walls of the cavity.

Split magnets, however, are useful for microtrons with superconducting cavities which have to be placed in regions of zero magnetic field.

In the split microtron it is possible to arrange the motion of the electrons such that at the various orbits compact electron bunches are formed at the nodes of the phase oscillations.[63] With the aid of these bunches, it was possible to excite oscillations with a wavelength down to 1 mm in an external cavity. However, as previously, there was no feedback between the radiated field and the electron bunch and thus, we cannot consider this scheme as a promising generator of millimeter waves.

Interest in the azimuthally nonuniform systems initially started in connection with the proton microtron project of

1 GeV proposed by Roberts.[45] This was essentially one of the first proton accelerator projects which are now known as meson factories. The primary difficulty of accelerating protons in a microtron is connected with its large rest energy. The increase in energy per revolution can be decreased as shown in Eqs. (1.5) and (1.10) by changing to an injection with large values of m (m ∿ 1000). In such a microtron, particles must perform each revolution during a large number of periods of the field T_0, the tolerances on the magnetic field and the frequency of the accelerating field are very high. This is one of the reasons why this project was not considered further.

All the above-mentioned microtrons are of the pulsed type. This is determined by the high frequency generators which could be used in microtrons and is not necessarily a characteristic of the accelerator itself. Contemporary development of high power electronics give the possibility of constructing a continuous wave (cw) microtron. Such an accelerator with beam power of several tens of kilowatts can be rightfully called an electron cyclotron.[17,18]

A cw microtron could be very useful for radioactivation analysis and for chemical technology as well as for the study and development of materials which have to withstand large doses of radiation. A cw microtron would also be of considerable significance for nuclear physics. Indeed, for neutron physics, the pulsed mode is useful only in work involving the time-of-flight methods. For experiments based on coincidence methods the pulsed mode increases the background of random coincidences and overloads the counter during a pulse. A cw microtron with a well collimated beam and a highly controllable energy will undoubtedly open new possibilities for studying photonuclear reactions and electron scattering at the intermediate energies. Classical sources of electrons for such purposes, betatrons and synchrotrons, have an intensity which is thousands or tens of thousands times less than that which can be expected from a cw microtron (see Chapter VIII).

The high stability of the energy and the highly collimated beam make the microtron very attractive as an injector into other accelerators, particularly synchrotrons. Beam parameters of some of the injector microtrons are given in Table 1.2. The energy spread and the emittance of the microtron beam is well suited to capture into a synchrotron. The changeover of synchrotrons to microtron injection which insures a higher energy for the injected particles, raises the current of accelerated electrons in the synchrotron because of two circumstances. First of all, the Coulomb forces are lower at injection; secondly, injection occurs in high magnetic fields where field distortion is smaller, and the gradient larger, making it possible to capture a larger number of particles. The experience at Frascati, where a 12-MeV microtron injector has been built, has led to a 20-fold increase in beam intensity with a marked increase in the stability of the whole system.[93]

The compatibility of the microtron with a strong focusing synchrotron was investigated by O. Wernholm.[59] The use of a microtron as an injector into a weak focusing synchrotron

TABLE 1.2 Parameters of microtrons, used as injectors.

	Particle Energy (MeV)	Number of Orbits	Beam Size (Vertical and Radial) (mm)	Angular Spread in mrad		Emittance mm·Rad	
				Vertical	Radial	Vertical	Radial
United Institute for Nuclear Research, Dubna, USSR	30	30	1.5 × 3	0.5	5	0.7×10^{-3}	1.5×10^{-2}
Physical Institute, Academy of Science, USSR, Moscow	7	10	2 × 4	1.5	15	3×10^{-3}	6×10^{-2}
Lund University, Sweden	6.4	10	7 × 7	2	7	1.4×10^{-2}	5×10^{-2}
University of Wisconsin, P.S.L.	44	34	2 × 4			1.5×10^{-2}	3×10^{-2}

and the stacking of particles in a synchrotron was reported in the reports of the Lebedev Physical Institute of the Academy of Sciences, USSR.[66,67]

Development of the microtron injector in the Lebedev Physical Institute led to the invention by K. A. Belovintsev and F. P. Denisov of the positron microtron[67,68]; this ingenious scheme allowing the acceleration of positrons in the microtron will be described in the next Section.

SECTION 5 - THE POSITRON MICROTRON

The initial scheme for the positron microtron is shown in Fig. 1.15. Electrons are initially accelerated in a flat

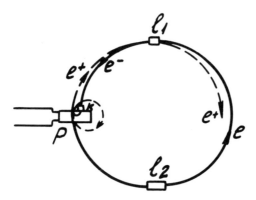

Fig. 1.15 Schematic of the positron microton.

cavity (acceleration of the first type); at the last orbit the electron beam is displaced by two short magnetic channels such that they strike the walls of the cavity opposite the emitter. At this point a tungsten electron positron converter about 1 mm thick (0.3 radiation lengths) is located behind

MICROTRON

which a light graphite or aluminum positron attenuator is placed.

The difference between the lengths of the magnetic channels is equal to the distance of the positron emission point to the axis of the cavity, while the sum of the lengths of the channels is equal to the phase difference which is needed for positron capture into the accelerating mode: $\ell_2 - \ell_1 = x_0$, $\ell_1 + \ell_2 = \pi$, 3π, etc. (all the lengths are expressed in units of $\lambda/2\pi$).

If these conditions are fulfilled, the positrons will be captured into an acceleration mode and will move in the opposite direction to the electrons. When the positrons reach the last orbit, the action of the magnetic channels displaces them such that they go through the positron extraction channel.

Detailed investigation into the conditions needed for electron-positron conversion and their capture into the acceleration mode can be found in reports by K. A. Belovintsev and P. A. Cherenkov.[6,7] Calculations show that accelerated electrons can be converted with an efficiency of 10^{-5}; subsequent measurements have shown that this coefficient is indeed $2-5 \times 10^{-6}$, in fairly close agreement with calculations.

Another design for a positron microtron, also built at the Lebedev Physical Institute, is shown in Fig. 1.16. Es-

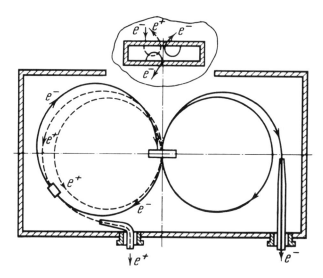

Fig. 1.16 The double microtron of the Physical Institute, Academy of Sciences, USSR.

sentially, this method involves a double microtron, the right-hand side of which accelerates electrons while the left side accelerates positrons. In this method, it is possible to expect a larger conversion coefficient, and it is possible to simultaneously extract an electron beam, a condition which is essential for parallel stacking of positrons and electrons in a storage ring system for which it was primarily designed. The second layout is somewhat more bulky than the first one. However, both methods permit effective acceleration of positrons. We must emphasize that the positron beam is just as compact and monoenergetic as the electron beam. One should keep this in mind when comparing a positron microtron with a linear accelerator in which positrons are converted and accelerated; in the latter case although the positron current in the beam is larger (conversion coefficient of 7×10^{-3} (see Ref. 88), the positrons have a much larger spread in energy and transverse momentum.

CHAPTER II

PHASE MOTION IN THE MICROTRON

SECTION 1 - SMALL PHASE OSCILLATIONS

As in other resonant cyclic accelerators, in the microtron the accelerating field ensures the phase stability of the particles. The energy and phase of the accelerated particles oscillate about equilibrium values. Phase stability is due to the fact that the motion is not isochronous, the dependence of the time of revolution on the energy through the relativistic change in mass being the principal action of the microtron that insures stability of the accelerated particles.

As has been pointed out, the microtron is an accelerator with a varying harmonic number and a large energy increase per revolution. These characteristic peculiarities which exist only in the microtron set it apart from other cyclic accelerators and determine the unique nature of its phase stability.

In the study of the microtron dynamics it is convenient to introduce[4] a number of dimensionless parameters that simplify the equations and give them a more general character. First of all, we will introduce the wave number

$$k = \frac{\omega}{c} = \frac{2\pi}{\lambda} , \qquad (2.1)$$

where ω and λ are the angular frequency and the wavelength of the high frequency field, and we will replace the coordinates and linear dimensions with dimensionless quantities

$$x = kX, \quad y = kY, \quad z = kZ, \quad \ell = kL, \quad \rho = kR, \qquad (2.2)$$

where X, Y, Z, are the Cartesian coordinates, L is the thickness of the cavity, R is the orbit radius (see Fig. 2.1), so that a unit of length is equal to $\lambda/2\pi$.

Next, we will substitute for time the phase ϕ

$$\phi = \omega t , \qquad (2.3)$$

and introduce dimensionless velocities

$$u = \frac{1}{c}\frac{dX}{dt} = \frac{dx}{d\phi} , \qquad v = \frac{1}{c}\frac{dY}{dt} = \frac{dy}{d\phi} \qquad (2.4)$$

$$w = \frac{1}{c}\frac{dZ}{dt} = \frac{dz}{d\phi} , \qquad \beta = \sqrt{u^2 + v^2 + w^2} \qquad (2.4)$$

normalizing them to the speed of light c.

Previously we introduced the relativistic factor (see Chapter I)

$$\Gamma = \frac{U}{U_o} = \frac{1}{\sqrt{1 - \beta^2}} \qquad (2.5)$$

where U is the particle energy and U_o its rest energy. In a similar way we introduce dimensionless transverse momentum, the vertical momentum p and the horizontal momentum q in units of mc:

$$p = \frac{\Gamma}{c}\frac{dZ}{dt} = \Gamma w$$

$$q = \frac{\Gamma}{c}\frac{dX}{dt} = \Gamma u . \qquad (2.6)$$

Finally, we introduce the parameter ε, which characterizes the electric field in the cavity; it is equal to

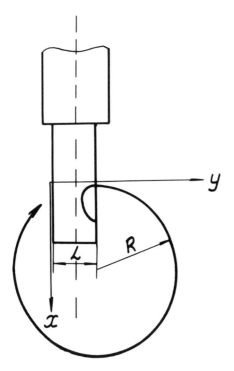

Fig. 2.1 Coordinate system.

$$\varepsilon = E/H \tag{2.7}$$

where E is the amplitude of the electric field at the axis of the cavity and H is the constant magnetic field at the accelerator. The magnetic field H also enters through the parameter

$$\Omega = H/H_o \tag{2.8}$$

introduced in Chapter I; $H_o = mc\omega/e$ -- H_o is the cyclotron field at frequency ω.

The dimensionless high frequency field in the cavity is proportional to the product $\varepsilon\Omega$, while the dimensionless orbit radius which is proportional to the pulse is determined by the equation

$$\rho = \frac{\beta\Gamma}{\Omega} ; \tag{2.9}$$

where $\Gamma \gg 1$, we have $\beta \approx 1$ and $\rho \approx \Gamma/\Omega$.

In the equations of phase motion in the microtron we will first consider the ideal case, one where the thickness of the accelerating gap is zero, i.e., the distribution of the accelerating field along the y axis is a δ-function. Such an approximation can be used after several initial revolutions where the orbit radius is considerably larger than the thickness of the cavity and the particle velocity is close to the speed of light. For an infinitely thin accelerator gap phase motion was initially treated in detail by A. A. Kolomensky.[46,47]

The dimensionless time of revolution along the circumference is equal to

$$\tau = \frac{2\pi\rho}{\beta} = \frac{2\pi\Gamma}{\Omega} , \tag{2.10}$$

and the equations describing the change in phase and energy after one revolution have the form

$$\phi_{n+1} = \phi_n + \frac{2\pi}{\Omega} \Gamma_n ,$$

$$\Gamma_{n+1} = \Gamma_n + A \cos \phi_{n+1} , \tag{2.11}$$

where ϕ_n is the phase in which the particle passes the accelerating gap, Γ_n is the resulting particle energy (see Fig. 2.2), and $A = eV/mc^2$ is the amplitude of the dimensionless potential in the gap.

From Eqs. (2.11) it is easy to determine the equilibrium values of energy Γ_s and phase ϕ_s. After passing the accelerating gap, let the equilibrium energy of the particle increase by a value $\Delta\Gamma_s$ such that the period of revolution increases by g periods of the accelerating field. Then for the values Γ_s and ϕ_s we have the following relationships:

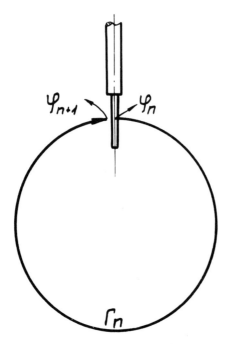

Fig. 2.2   For calculating phase oscillations.

$\Delta \Gamma_s = \Gamma_{s,n+1} - \Gamma_{s,n} = A \cos \phi_s = g\Omega$,

$\Gamma_{s,n} = \Omega[m + g(n - 1)]$,

$\phi_{s,n} = \phi_s + 2\pi(n - 1)\left[m + \frac{g}{2}(n - 2)\right]$. (2.12)

It is interesting to note that the case g = 0 corresponds to conventional cyclotron acceleration (a single acceleration per revolution) where

$\cos \phi_s = 0$, $\phi_s = \pi/2$, $\Gamma_{s,n} = m\Omega = $ const.

These relationships indicate that stationary resonance acceleration is not possible in a cyclotron. Indeed, particles in a cyclotron are accelerated in the first half period of the phase oscillation and if phase motion at that time is not terminated by extraction, the particles move into a decelerating field and are consequently slowed down. In a microtron g ≠ 0 and energy increases in proportion to n, i.e., a resonance acceleration of particles exists and the harmonic order of acceleration increases in proportion to the energy of the particles.

Assuming the deviation of the phase and the energy of the particles from the equilibrium value

$$\eta_n = \phi_n - \phi_{s,n}, \qquad \gamma_n = \frac{2\pi}{\Omega}(\Gamma_n - \Gamma_{s,n}) \qquad (2.13)$$

as being small we can linearize the system of Eqs. (2.11) and present them as

$$\begin{pmatrix} \eta_{n+1} \\ \gamma_{n+1} \end{pmatrix} = \begin{pmatrix} 1 & 1 \\ -2\pi g \tan \phi_s & 1 - 2\pi g \tan \phi_s \end{pmatrix} \begin{pmatrix} \eta_n \\ \gamma_n \end{pmatrix} \qquad (2.14)$$

In the linear approximation the phase motion of the particles is expressed by such a matrix equation. If ϕ_s = const, then the matrix elements entering into Eq. (2.14) are constant and independent of the number of revolutions. The values η_n and γ_n corresponding to the nth orbit can be expressed through the initial values η_0 and γ_0 with the aid of this matrix raised to the nth power. In fact the motion is expressed through matrix elements with well known relationships (see, for example, Ref. 48, and a solution of these relationships given in Appendix II).

The main characteristic determining the behavior of motion is the trace of the matrix

$$\begin{pmatrix} \alpha_{11} & \alpha_{12} \\ \alpha_{21} & \alpha_{22} \end{pmatrix}$$

The trace is equal to $S = \alpha_{11} + \alpha_{22}$; its magnitude is invariant as regards the choice of the corresponding trajectory points at which the values are calculated (for instance the start or the end of the nth orbit). In this case we have

$$S = 2(1 - \pi g \tan \phi_s) \qquad (2.15)$$

The other main characteristic of motion is the determinant of the matrix (α_{ik}). Its magnitude is also invariant as regards the choice of the corresponding points. In this case, and from there on, the determinant of the matrix is equal to unity inasmuch as we investigate particle motion in a given field, neglecting radiation and dissipative effects and making use of canonically conjugate variables for which phase space is conserved.

Motion is stable as long as S is within the limits

$$-2 < S < 2, \qquad (2.16)$$

which corresponds to an equilibrium phase ϕ_s within the limits

$$0 < \phi_s < \arctan \frac{2}{\pi g} \qquad (2.17)$$

If ϕ_s falls within these limits, then the phase of the particle oscillates close to ϕ_s and in the linear approximation these oscillations are stable. Equations (2.14) and (II.7) (see Appendix II) lead to the expressions

$$\eta_n = - \frac{C}{\sqrt{2\pi g \tan \phi_s}} \cos(\nu n + \chi + \zeta),$$

$$\gamma_n = C \cos(\nu n + \chi), \qquad (2.18)$$

where

$$\cos \zeta = \sqrt{\frac{\pi g \tan \phi_s}{2}} \quad (0 < \zeta < \pi),$$

and C and χ are arbitrary constants determined by the initial conditions; the frequency ν is determined from

$$\cos \nu = \frac{S}{2} = 1 - \pi g \tan \phi_s. \qquad (2.19)$$

Outside the stable region (2.16), particle oscillation becomes complex and phase motion becomes unstable. For $S > 2$, the values η_n and γ_n increase exponentially (see Fig. 2.3).

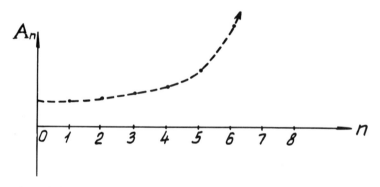

Fig. 2.3 Phase oscillations for $S > 2$ (η_n or γ_n expressed through A_n).

As can be seen from Eq. (2.17), this corresponds to values of $\phi_s < 0$, i.e., when the particles pass through the accelerating field rising in time. Such a region of instability exists in all cyclic accelerators.

Phase motion instability arising from $S < -2$, i.e., when

$$\phi_s > \arctan \frac{2}{\pi g}, \qquad (2.20)$$

is unique only of the microtron. In this case (Fig. 2.4), the oscillations change signs at each revolution and their amplitude increases exponentially.

This instability happens because the accelerating field forces leading to phase stability are too strong, and this

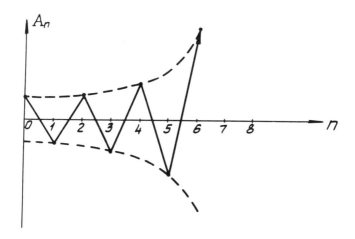

Fig. 2.4 Phase oscillation for $S < -2$.

leads to overfocusing and instability. Phase motion in other accelerators does not have instabilities of this type because the energy gain per revolution in them is too small and the motion for any value of ϕ_s occurs close to the boundary $S = 2$.

Utilizing these equations, let us now investigate certain properties of small phase oscillations. Let us first see how these properties of motion are related to the value g which determines the change in the harmonic number of acceleration from one revolution to the other.

When $g = 1$, the region of stable equilibrium phase is the widest; it is described by the inequality

$$0 < \phi_s < \arctan \frac{2}{\pi} = 32° \quad . \tag{2.21}$$

As g increases, this region rapidly becomes correspondingly narrower. Numerical examples presented in the following paragraphs show that the largest amplitude for which stable phase oscillations still exist is approximately equal to the width of the stable phase region. Thus, for systems with large values of g, the number of particles captured into the accelerating mode decreases. Besides, these systems require that the tolerance on the value of ϕ_s be much smaller, i.e., the tolerances on the accelerator parameters become more stringent. Finally, as can be seen from Eq. (2.12), it is either necessary to increase the amplitude of the potential A which is already large, or to decrease the magnetic field in proportion to the parameter Ω. All of these considerations indicate that from the practical standpoint, it is best to use accelerating modes for which

$$g = 1 \quad . \tag{2.22}$$

In all the following discussions, we will assume this value for g.

The range of ϕ_s (2.21), as we have already pointed out, is specific for the microtron; in other cyclic accelerators, the equilibrium phase can have any value from 0 to 90°. It is natural to expect that the optimum conditions for phase motion correspond to values of ϕ_s which lie essentially in the middle of the interval given by Eq. (2.21). If we remove from Eq. (2.18) the values sin νn and cos νn, we find the trajectory equation describing a point on the phase plane η_n, γ_n in the form

$$\eta_n^2 + \eta_n\gamma_n + \frac{1}{2-S}\gamma_n^2 = \frac{2+S}{4}\eta_{max}^2 = \frac{1}{4}\frac{2+S}{2-S}\gamma_{max}^2,$$
(2.23)

where η_{max} and γ_{max} are maximum possible values of η_n and γ_n for the given initial conditions.

The phase trajectory is an ellipse, the shape of which is determined by the value S (i.e., eventually by the equilibrium phase ϕ_s) and the size by the initial value of η and γ. For arbitrary initial conditions the amplitude of linear phase oscillations are bigger, the closer the ϕ_s approaches the boundary of the stable region, where $|S| = 2$. The amplitude of oscillations is smallest near the middle of interval (2.21) (for $S \approx 0$); the equilibrium phase which corresponds to the value $S = 0$ is

$$\phi_s = \arctan\frac{1}{\pi} = 18°$$
(2.24)

for which, according to Eq. (2.19), $\nu = \pi/2$, i.e., one full phase oscillation takes place during four revolutions. Oscillations which correspond to this case are shown in Fig. 2.5.

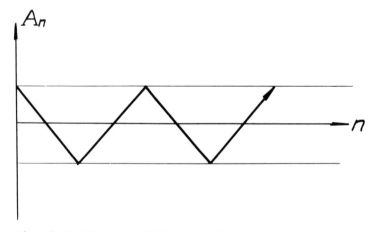

Fig. 2.5 Phase oscillations for S = 0.

MICROTRON

As follows from the theory described by the second order matrices (see Appendix II) for equilibrium phases lying close to the value (2.24), phase motion in the linear approximation is not seriously affected by various disturbing factors and thus, from the practical point of view, the equilibrium phases which lie close to the value (2.24) are optimal.

The phase oscillations in a microtron are specific in a number of respects. First of all, their period consists of 4 or 5 revolutions, whereas, in other cyclic accelerators, it is equal to hundreds or thousands of revolutions. Another distinguishing property is that the period of the phase oscillations, expressed by the number of revolutions, and their amplitude does not change with increase in the particle energy. In other words, the microtron does not have adiabatic damping of the oscillations. The increase in the longitudinal momentum of the particles, which in other accelerators causes damping of the phase motion, in the microtron is compensated by the increase in the phase length of the revolutions. As we will see later, the very same reason accounts for the absence of damping of vertical motion in the microtron. If these oscillations were damped, the effect of various occasional disturbances would have been weaker. However, the number of orbits in the microtron is not large, and for this reason, as experience has shown, it is possible to accelerate particles without significant losses.

The next and perhaps the most important property of phase dynamics in the microtron is the high monoenergeticity of the beam of accelerated particles. Phase oscillations are described by the variables η, γ, where γ according to Eq. (2.13) is the difference between the energy Γ and the equilibrium value, normalized to the parameter Ω. As we have seen above, the amplitude of phase oscillations in these variables does not change. Accordingly, the absolute energy spread in the beam is proportional to the parameter Ω and does not change with increase in energy. Inasmuch as the energy gain per revolution is also proportional to Ω, the relative energy spread does not depend on Ω and goes as $1/n$, where n is the number of orbits. In this way, the monoenergeticity of the accelerated beam grows with increase in the number of orbits.

A quantitative evaluation of the energy spread will be given in Section 3 based on numerical calculations of the nonlinear phase oscillations. At present we note that the relative energy spread of the beam in the microtron is much smaller than in the linear accelerator. This property of the microtron is very important when it comes to using it in many areas, particularly in nuclear physics and as an injector into other accelerators.

In conclusion we will point out two more properties of phase oscillations. As can be seen from Eqs. (2.14) and (2.19), all the characteristics of the oscillations are determined by the equilibrium phase ϕ_s. Other accelerator parameters, for instance the strength of the magnetic field or the amplitude of the accelerating potential, enter into the equations in a combination expressed through the tangent of ϕ_s. The velocity of the particles also does not enter the expression describing phase motion. Thus, it is independent of whether the accelerated particles are relativistic or not.

Both indicated properties depend on the fact that we investigated an idealized system in which the accelerating gap is considered to be zero. As we will see from Sections 3 and 4, taking into account a finite thickness cavity changes things, and the phase motion on the first orbits is determined not only by the equilibrium phase (2.12) but by other parameters also. For this reason particle velocity on the first orbit cannot be too low, although it could be noticeably less than the velocity of light.

SECTION 2 - NUMERICAL CALCULATIONS OF NONLINEAR PHASE OSCILLATIONS

For further investigation of phase motion it is convenient to delete the value of Γ_n from the analysis. This can be accomplished by writing (2.11) for two consecutive cycles n, and (n + 1), expressing Γ_n and Γ_{n+1} through ϕ_n, ϕ_{n+1} and ϕ_{n+2}. In this way we obtain the difference equation of the second order

$$\phi_{n+2} = 2\phi_{n+1} + \frac{2\pi}{\Omega} A \cos \phi_{n+1} - \phi_n \ . \qquad (2.25)$$

For small deviations from the equilibrium phase this equation can be linearized. The linear equation obtained can be easily solved (see Appendix II) and from it follow all the derivations concerning small oscillations which were mentioned above. If, however, the amplitude of the oscillations is so great that the nonlinear members cannot be neglected, then the full equation (2.25) does not have an analytic solution and must be solved numerically, since the values ϕ_0 and ϕ_1 are determined by the initial conditions.

First of all, let us point out that the nonlinear phase oscillations are not attenuated for apparently the same reasons as the small (linear) oscillations. This can be easily proved independently, by calculating the Jacobian of the transformation from the variables ϕ_n, Γ_n to the variables ϕ_{n+1}, Γ_{n+1}. From Eqs. (2.11) it is equal to unity.

Before we present the results of the calculations we will transform Eq. (2.25) by replacing the dimensionless time parameter ϕ equal to the total phase build-up by the phase oscillation parameter ψ which is equal to ϕ subtracted multiple of 2π. We will also introduce the first and second order differences by the equations

$$\Delta a_n = a_{n+1} - a_n \ , \qquad \Delta^2 a_n = \Delta a_{n+1} - \Delta a_n \ , \qquad (2.26)$$

where a_n is the arbitrary function of the index n. Taking into account the third equation in (2.12) we have equations

$$\Delta^2 \phi_n = \Delta^2 \phi_{s,n} + \Delta^2 \psi_n = 2\pi g + \Delta^2 \psi_n \ . \qquad (2.27)$$

Thus, we can rewrite Eq. (2.25) in the form

$$\Delta^2 \psi_n = \frac{2\pi}{\Omega} A \cos \psi_{n+1} - 2\pi g \qquad (2.28)$$

or, by making use of the first equation in (2.12) in the form

$$\Delta^2 \psi_n + 2\pi g \left(1 - \frac{\cos \psi_{n+1}}{\cos \phi_s} \right) = 0 \quad . \qquad (2.29)$$

If we replace the increment $\Delta^2 \psi$ by the differential $d^2 \psi$, then Eq. (2.29) transforms to that of a pendulum with an external torque and will be equivalent to the equation which is used to describe phase motion in other cyclic accelerators. However, the transition from a *difference* equation to a *differential* equation in the case of a microtron, generally speaking, may not hold. It can only be justified in the case when the frequency of phase oscillations is considerably smaller than the revolution frequency.

Strictly speaking, phase motion in all cyclic accelerators is described by the difference equations since the accelerating field is localized in one or several accelerating gaps. However, the energy gain of the particles per revolution usually is many times smaller than their total energy. Thus, the frequency of phase oscillations is small and it is possible, with a fair degree of accuracy, to replace the difference equation by the differential equation.

In this case, the accelerator field is, so to speak, spread out over the whole revolution. Mathematically, this means that we take into account only the azimuthal harmonic of the accelerating field which corresponds to the harmonic order of acceleration. The phase velocity of this harmonic is close to the velocity of the particle and this resonance harmonic accelerates the particles. In the microtron the harmonic number changes from orbit to orbit and thus, this procedure cannot be used.

The equation of small phase oscillations in the microtron can be easily obtained in another way, namely, by linearizing Eq. (2.29). Then, for the quantities $\eta_n = \psi_n - \phi_s$, we obtain the linear equation

$$\Delta^2 \eta_n + (2 - S)\eta_n = 0 \quad . \qquad (2.30)$$

For $\phi_s \approx 0$, we have $S \approx 2$ and the frequency of oscillations is small, so that the difference equation can be replaced by the differential equation. However, in this case, the phase oscillations for the smallest noticeable amplitude appear to be substantially nonlinear, and thus, Eq. (2.30) cannot be used for them Aside from that, this case is of no interest from the practical standpoint.

In practice it is important to study phase motion for the condition in Eq. (2.24). This was accomplished by means of numerical analysis by A. A. Kolomensky.[47] The results of these calculations are shown in Fig. 2.6. This figure also shows a family of curves representing this solution of the differential equation obtained by the substitution of $d^2\psi$ for $\Delta^2\psi$ (family 1). The motion in the phase plane for some initial phases is shown in Fig. 2.7.

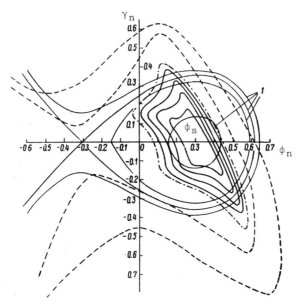

Fig. 2.6 Nonlinear phase oscillations in a microtron (Ref. 47).

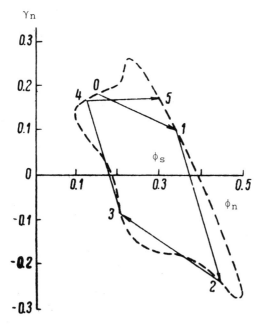

Fig. 2.7 Motion of a representative point on the phase plane for S = 0.

MICROTRON 35

Phase trajectories have a fairly complicated form and differ noticeably from the ellipse (2.23), even when $\eta_{max} \sim 0.15$. This means that even at this amplitude, oscillations have already a substantially nonlinear character.[99] Oscillations become unstable (dotted lines on Fig. 2.6) when their amplitude exceeds the maximum allowable; the latter is approximately equal to the size of the region of stable equilibrium phases. The shape of the phase trajectories differs substantially from the shape of the curves obtained from the solution of the differential equation. It is interesting to note that it was recently demonstrated by L. G. Lugansky that for $32° < \phi_s < 35.5°$ it is possible to have stable phase oscillations.[40] The region of phase stability in this case is doubly connected and the phase skips from one region to another (see also Section 2, Chapter VII, and Ref. 94).

SECTION 3 - CONDITIONS OF CAPTURE AND RESONANCE ACCELERATION OF PARTICLES

The resonance acceleration in a microtron is possible when two conditions are fulfilled. First of all, the energy gained per revolution must correspond to condition (1.3) which leads to the first relationship (2.12). Secondly, the particle energy at the first revolution (n = 1) must correspond to the second equation of (2.12) and their phase must lie in the vicinity of the equilibrium phase. Now, knowing the properties of phase motion we can state these conditions more accurately.

We will first investigate the first condition. For g = 1, it can be rewritten in the form

$$\Delta \Gamma_s = \Omega \; , \qquad (2.31)$$

while the first equation in (2.12) gives

$$\cos \phi_s = \frac{\Omega}{A} \; . \qquad (2.32)$$

As we have already seen, in order to have the microtron mode of acceleration, it is necessary that the equilibrium phase ϕ_s lie in a region of stability (2.21) and, besides, it must lie close to the optimum phase (2.24).

For a finite thickness of the accelerating cavity, the energy gain depends on the velocity of the particle, i.e., on their energy and number of orbits. After several revolutions the velocity of the particles is very close to the speed of light, and the orbit radius is considerably larger than the thickness of the cavity so that the curvature of the orbit inside the cavity can be neglected. Thus, on all the subsequent orbits the energy gain is identical. It is generally required that this energy gain satisfies (2.31).

Let us now calculate the energy gain. In flat cavities which are used in most modern microtrons, the electric field is constant along the axis of the cavity (the y axis on Fig. 2.1) and depends on the transverse coordinate

$$E_y = EF_0(x) \cos \phi \quad , \tag{2.33}$$

where the function $F_0(x)$ is determined by the shape of the accelerating cavity. Proceeding according to Eqs. (2.4), $u = w = 0$ and $v = \beta = 1$, we will write the equations which determine the energy change

$$\frac{d\Gamma}{d\phi} = \varepsilon \Omega F_0(x) \cos \phi \quad . \tag{2.34}$$

By solving it and then substituting it into (2.31), we obtain the main condition for acceleration in the form (see Ref. 4)

$$\frac{\Delta \Gamma_s}{\Omega} = 2\varepsilon F_0(x) \sin \frac{\ell}{2} \cos \phi_s = 1 \quad , \tag{2.35}$$

where ℓ is the thickness of the cavity and x is the radial position of the beam.

Equation (2.34) shows that the dimensionless amplitude of the potential in the cavity is equal to

$$A = \varepsilon \Omega F_0(x) \ell \quad , \tag{2.36}$$

and its effective value which determines the energy gain according to (2.35) is equal to

$$A_{eff} = A \frac{\sin \ell/2}{\ell/2} \quad , \tag{2.37}$$

i.e., it is equal to the product of A and the "transit factor" generally used in the theory of klystrons and other high frequency devices.

We see that the energy gain per revolution on the outer orbits is constant and (for the same amplitude potential) differs from the gain at the gap of zero thickness by the constant factor. Thus, at the outer orbits, the cavity can be replaced by an equivalent gap of zero thickness and an effective potential A_{eff}.

Let us point out, however, that the field in flat cavities changes radially and thus, the energy gain depends on the radial coordinate of the beam. In Eq. (2.35) this is determined by the function $F_0(x)$ which must be taken into account in numerical calculations.

A particle can be accelerated in a stable mode in that case where the amplitude of phase oscillation is not very great; that is, it does not exceed the maximum allowable amplitude. This satisfies the second condition for resonance acceleration.

In order to determine the maximum amplitude in a real cavity of finite thickness, the equation of particle motion was integrated numerically on a computer.[4] In Fig. 2.8 phase motion is shown for the first type of acceleration in a cavity of thickness $\ell = 1.2$, obtained by numerical calculations. We can see that the initial conditions (which depend on the particle exit phase from the emitter and its position) determine the character of motion only in the first several orbits;

MICROTRON

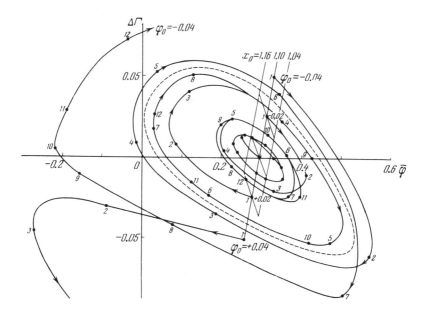

Fig. 2.8 Phase oscillations and the separatrix.

on the outer orbits the character of motion is determined by the phase space available for phase oscillations which we investigated above.

In Fig. 2.8 it can also be seen that the general rules governing phase motion in cavities of finite and zero thicknesses are identical (see Figs. 2.6 and 2.8). For small amplitudes, phase oscillations in a microtron with a finite thickness cavity are stable; in general, calculations for the first twelve orbits did not uncover any tendency for oscillation growth. When, however, the amplitude of oscillations exceeds a certain predetermined value, then the character of motion changes drastically, and it becomes unstable. The limiting phase trajectory which separates stable motion from unstable motion will be called the separatrix, a term used in the analysis of nonlinear oscillations described by differential equations.

However, we must point out that phase motion close to the separatrix in the microtron differs substantially from other cyclic accelerators. In other accelerators motion on the separatrix has a limiting character, i.e., the period of motion goes to infinity; this is analogous to the swinging of a pendulum where the kinetic energy at the highest point becomes equal to zero. In the microtron, however, the opposite is true, the phase period changes slightly in crossing over the separatrix, but the oscillations begin to grow rapidly; this is a unique property of phase motion described by the difference equation.

In the general theory of nonlinear oscillations, the separatrix is the result of the exact analytical solution of a differential equation and thus, it is known that it is indeed a continuous curve separating the phase plane into regions in which phase trajectories have different topological characteristics. The separatrices in Figs. 2.6 and 2.8 are the result of numerical solutions. Their positions are only known approximately and, in general, it is not clear whether the separatrix exists as a total determinable continuous curve. It may be that the properties of phase motion change drastically in some finite amplitude interval. Strictly speaking, it is not even known whether the motion inside the separatrix is indeed stable or unstable, but the instability is so weak that it cannot be noticed in a small number of revolutions.*

All these questions are interesting from the theoretical point of view. For practical reasons, the stability of the process of acceleration for an arbitrary number of orbits is not essential, what is important is the *dynamic aperture* of the accelerator, i.e., the region of initial parameters for which electrons are accelerated for a predetermined number of orbits. This region is obtained for a limited number of orbits with the aid of numerical calculations and is adequate from a practical standpoint for evaluating particle capture.

It is interesting to note that numerical calculations performed by us for finite thickness cavities gave phase trajectories and separatrices which appeared continuous and closely resembled ellipses. Calculations performed by A. A. Kolomensky gave phase trajectories which substantially differed from ellipses. Phase trajectories and separatrices of somewhat unusual shape were also obtained numerically[49] and graphically[55] on the basis of Eq. (2.11) for infinitely thin cavities. For cavities of finite thickness, there takes place an averaging process which makes the curves become smoother.

Thus we arrive at the second condition for resonance acceleration in a microtron: the point representing the particle in phase space must fall within the region of phase stability bounded by the separatrix. The initial position of this point must correspond to an orbit for which, on further acceleration, the shape of its phase trajectory remains unchanged, or changes very slightly. This condition of resonance acceleration was used in work reported in Ref. 22.

As an example, on Fig. 2.9 is shown phase motion for the second type of acceleration in a cavity of thickness $\ell = 1.45$. During the first two revolutions the energy gain per revolution changes noticeably and the value of the average phase is relatively indeterminant since it is not clear how to calculate it when the orbit curvature is comparable to the

---

*A detailed study of phase stability by Melekhin has led to the discovery of nonlinear resonances that limit the number of orbits in a microtron.[96,99]

MICROTRON 39

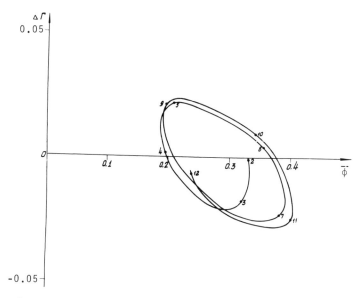

Fig. 2.9 Phase trajectory for a finite thickness accelerating cavity.

thickness of the cavity. Thus, at its very start, the phase trajectory appears to have a rather unusual shape having very little in common with an ellipse. However, on further acceleration the energy gain becomes constant, the average phase of flight is determined by a single quantity, and the phase trajectories take on a familiar shape.

The results presented allow us to evaluate the nonmonochromaticity of the accelerated beam in the microtron. The maximum possible energy spread in the beam is determined by the size of the separatrices and in dimensionless units amounts to ± 0.05 Ω [see Fig. 2.8 and also Eq. (3.21)]. By taking into account the second equation in (2.12) the corresponding energy spread will be equal to ± 0.05/n, where n is the number of orbits.

SECTION 4 - THE DEPENDENCE OF PHASE MOTION ON THE THICKNESS OF THE ACCELERATING CAVITY

The effect of the thickness of the cavity on phase motion has been investigated in the previous section. At this time we will discuss this problem in more detail following the work reported in Ref. 14.

It was shown above that on the outer orbits where the particle velocity is almost equal to the speed of light and the orbit radius is considerably greater than the thickness of the cavity, the real cavity can be replaced by some equivalent gap of zero thickness. Thus, on the outer orbits, phase motion can be completely determined by the relationship

Γ/Ω. It follows from this that for proportional changes in the total particle energy and the magnetic field, phase motion does not change. If this were true in the initial orbits, then it would have been possible to accelerate nonrelativistic particles of low kinetic energy in the weak magnetic field. It appears, however, that the effect of the finite thickness of the gap in the cavity which exists in the initial orbits determines the lower limit of the accelerated particle energy.

Let us investigate this problem with the idea, not so much of obtaining exact quantitative results (they can be obtained easily by numerical integration of the equation of motion), but of explaining the physical aspect of the existing limitations.

When we consider a finite thickness cavity, equations of phase motion (2.11) change, and their nature depends on the determination of the phase ϕ_n. From here on we will assume that the phase ϕ_n corresponds to the moment when the particle crosses the midplane of the cavity (the plane containing the point 0 in Fig. 2.10). The nomenclature here is

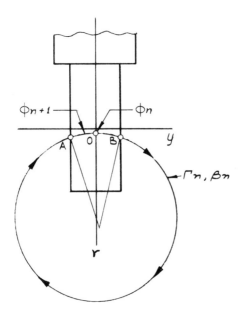

Fig. 2.10 Accounting for a finite thickness accelerating gap.

the same as previously, i.e., β_n and Γ_n is the velocity and the energy of the particles outside the cavity after passing through it at the phase ϕ_n. Let the average velocity of the particle in region OB at the exit from the cavity be equal to

MICROTRON 41

$\beta_n - \Delta\beta_1$, and on re-entering the cavity in region AO the average velocity is $\beta_n + \Delta\beta_2$. Let us designate the dimensionless length of these regions S_{OB} and S_{AO}. Then in place of the first equation in (2.11), we obtain the relationship

$$\phi_{n+1} = \phi_n + \frac{2\pi}{\Omega} \Gamma_n \delta\phi_n$$

$$\delta\phi_n = \left(\frac{1}{\beta_n - \Delta\beta_1} - \frac{1}{\beta_n}\right) S_{OB} - \left(\frac{1}{\beta_n} - \frac{1}{\beta_n + \Delta\beta_2}\right) S_{AO}$$

(2.38)

Additional terms in this expression come from taking into account the change in velocity during motion inside the cavity. Assuming that in the first approximation

$$S_{OB} = S_{AO} = \frac{S_{AB}}{2}, \qquad \Delta\beta_1 = \Delta\beta_2 = \frac{\Delta\beta_n}{4},$$

where $\Delta\beta_n$ is the total change in velocity after passage of the particle through the cavity, we obtain

$$\delta\phi_n = \frac{S_{AB}}{16} \frac{(\Delta\beta_n)^2}{\beta_n^3} = \frac{\ell}{16} \frac{(\Delta\Gamma)^2}{\Gamma^6}$$

(2.39)

For small values of Ω we have $\Gamma \gtrsim 1$ and $\delta\phi_n \sim \Omega^2$, and for larger values of Ω used in evaluating $\Gamma \approx (m + n)\Omega$ and $\delta\phi_n \sim (m + n)^{-6} \Omega^{-4}$, where m represents phase length of the first orbit ($m \geq 2$). In all cases the value $\delta\phi_n$ is very small and quickly diminishes as the number of orbits increases.

Taking into account the difference of S_{AO} from S_{OB} or the influence of the high frequency field on the shape of the trajectory leads to other additional terms which can also be neglected because of their small size. In this way, with the indicated choice of phase ϕ_n the first equation in the system (2.11) practically does not change at all. The second equation in the system (2.11) changes more significantly when we take into account the finite thickness of the cavity. The particle energy gain during passage through the cavity is determined by the equation

$$\frac{d\Gamma}{d\phi} = \varepsilon\Omega F_0(x) \cos\phi \frac{dy}{d\phi} .$$

In the first approximation one can consider

$$\frac{dy}{d\phi} = \bar{\beta} = \text{const},$$

then the energy gain is equal to

$$\Delta\Gamma_n = \varepsilon\Omega F_0(x) \ell \frac{\sin \Delta\phi_n/2}{\Delta\phi_n/2} \cos\phi_n ,$$

(2.40)

where $\Delta\phi_n = S_{AB}/\beta_n$, which is the transit time. In expression (2.40), the curvature of the orbit and the variation of β from unity are taken into account, thus it is more accurate than (2.35).

As the particle energy increases, the transit time $\Delta\phi_n$ becomes smaller from the decrease of S_{AB} (as the trajectory becomes straighter), as well as the increase in $\bar{\beta}$. With this, the transit factor increases approaching the limiting value $\frac{\sin \ell/2}{\ell/2}$; at the same time there is a corresponding increase in the energy gain per revolution. It is clear that due to this disturbance in phase motion a certain portion of the particles will leave the stable phase region and will be lost.

If the energy gain changes relatively slowly, then it is possible to introduce an effective amplitude potential at the nth orbit

$$A_n = \varepsilon\Omega F_0 \ell \frac{\sin \Delta\phi_n/2}{\Delta\phi_n/2} \qquad (2.41)$$

and the alternating equilibrium phase

$$\phi_s^{(n)} = \arccos \frac{\Omega}{A_n} \qquad (2.42)$$

as a result of which phase oscillations will take place [see Eqs. (2.32), (2.36), and (2.37)].

Accelerator parameters are selected such that at the outer orbits the value $\phi_s^{(n)}$ lies in the center of the stable region, i.e., that it be determined by Eq. (2.24). Thus, for low particle velocities at the inner orbits, the phase $\phi_s^{(n)}$ appears to be small, the stable region is narrow, and the number of captured particles is not significant. This accounts for the limits on the drift of the equilibrium phase $\phi_s^{(n)}$ and the determination of the minimum possible particle velocity at the first revolution.

Let us determine these values. From Eq. (2.41) it follows that the change in the transit time $\Delta\phi$ for $\Delta\phi < 1$ leads to the following change in the effective amplitude

$$\frac{\delta A}{A} \approx -\frac{\Delta\phi}{12} \delta(\Delta\phi) \quad . \qquad (2.43)$$

Numerical calculation shows that on the first revolution, particles leave the cavity almost parallel to its axis. In such a case.

$$\Delta\phi = \frac{S_{AB}}{\bar{\beta}_1} \approx \frac{\ell}{\bar{\beta}_1}\left[1 + \frac{1}{6}\left(\frac{\ell}{\rho_1}\right)^2\right] \quad , \qquad (2.44)$$

where $\rho_1 = \beta_1\Gamma_1/\Omega$ is the radius of the first orbit. Assuming that the limiting drift in the equilibrium phase is equal to approximately 10°, then from Eq. (2.42) we obtain the corresponding limiting change in the effective amplitude to be about 4%. Substituting this value into Eq. (2.43), determining the limiting value $\delta(\Delta\phi)$ and then using Eq. (2.44), we determine that for a thickness $\ell = 1.0-1.2$ and $m = 2$, i.e., for $\Gamma_1 = 2\Omega$, the minimum allowable velocity on the first revolution is $\beta_{1min} = 0.6-0.7$. Further lowering of the magnetic field is possible in acceleration modes for which $\Gamma_1 = 3\Omega$, 4Ω, ... In these modes, for the same values of velocity β_{1min}, the corresponding values of the parameter Ω are lower. (These modes have been considered in detail in Ref. 95.)

In obtaining these results we ignored the change in the accelerating field along the radius which is determined by the function $F_c(x)$. This is justified if the orbits are close to the axis of the cavity. If, however, these orbits are far from the axis where the dependence of the field on the radius is much greater, then straightening the trajectory will lead to additional changes in A_n which are not accounted for in these equations. In that case, the value β_{1min} can increase or decrease somewhat, depending on which side of the axis the orbits pass. This must be taken into account in calculating β_{1min} for orbits that are far from the axis.

In conclusion, let us note that the value Ω_{min} achieved in our first microtron[14] was equal approximately to 0.65, which is in close agreement with the evaluations presented above.

CHAPTER III

PARTICLE INJECTION CALCULATIONS

SECTION 1 - CONDITIONS FOR INJECTION

As we have already established in Chapter II [see Eq. (2.12)], the equilibrium value of energy gain per revolution $\Delta\Gamma_s$ and the total energy $\Gamma_{s,1}$ of the particle at the first orbit encircling the cavity are related to the parameter Ω which is proportional to the magnetic field

$$\Delta\Gamma_s = g\Omega \, , \quad \Gamma_{s,1} = m\Omega \, . \tag{3.1}$$

The value of $\Gamma_{s,1}$ depends on the method of injection. One of the simplest methods of injection consists of inserting the electrons into the cavity in the vicinity of its axis with some small initial energy $\delta\Gamma$ (for instance, with the aid of an electron gun). Particle energy on the first passage through the cavity increases by the same value $\Delta\Gamma_s$ as in subsequent revolutions; thus, the value of $\Gamma_{s,1}$ is equal to

$$\Gamma_{s,1} = 1 + \Delta\Gamma_s + \delta\Gamma \, . \tag{3.2}$$

Equating (3.1) and (3.2) we get

$$\frac{\Delta\Gamma_s}{g} = \Omega \, \frac{1 + \delta\Gamma}{m - g} \, . \tag{3.3}$$

If electron emission takes place from the walls of the cavity under the action of the accelerating field then $\delta\Gamma = 0$ (particle trajectory in this case is shown schematically in Fig. 3.1) and Eq. (3.3) takes on the form

$$\frac{\Delta\Gamma_s}{g} = \Omega = \frac{1}{m - g} \, . \tag{3.4}$$

As above, $g = 1$. From Eq. (3.4) it can be seen that the magnetic field is maximum when $m = 2$, i.e., if the first revolution around the cavity is twice as long as the period of the accelerating field. Then

$$\Delta\Gamma_s = \Omega = 1 \, . \tag{3.5}$$

This acceleration mode was used in the first microtrons with

MICROTRON

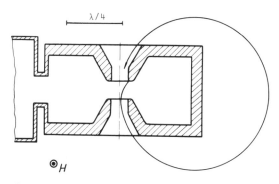

Fig. 3.1 Particle injection into a toroidal cavity.

toroidal cavities.

From the expression (3.5) two disadvantages which are unique to this method of injection can be seen. First of all, the magnetic field in the microtron is equal to the cyclotron value and this is comparatively small. Secondly, the energy of the accelerated particles can only take on values which are multiples of the rest energy with the result that for a fixed number of orbits, the energy in general cannot be changed.

As we mentioned in Chapter I, injection can be accomplished with the aid of an electron gun placed at the entrance of the cavity. This method makes it possible to increase the current of the accelerated particles. However, the magnetic field is small and also has a fixed value ($\Omega = 1 + \delta\Gamma \approx 1.15$).

In Chapter I we described injection methods in which electrons were initially accelerated in a cavity with a distributed electric field. The energy gain on the first half-revolution inside the cavity is designated through $\delta\Gamma^{(1)}$ and $\delta\Gamma^{(2)}$ (for the first and second types of acceleration respectively). For the first type of acceleration (see Fig. 1.7) we have equation

$$\Gamma_{s,1} = 1 + \delta\Gamma^{(1)} \tag{3.6}$$

and from Eq. (3.1) we obtain

$$\Delta\Gamma_s = \Omega = \frac{1 + \delta\Gamma^{(1)}}{m} \tag{3.7}$$

and for the second type of acceleration (see Fig. 1.7)

$$\Gamma_{s,1} = 1 + \Delta\Gamma_s + \delta\Gamma^{(2)} \tag{3.8}$$

$$\Delta\Gamma_s = \Omega = \frac{1 + \delta\Gamma^{(2)}}{m - 1} \tag{3.9}$$

(we consider g as always being equal to unity).

Numerical calculations presented in the following paragraphs show that for these methods of injection the parameter Ω can vary within wide limits; it can either be greater or less than unity and the energy of accelerated particles changes correspondingly. By this means it is possible to remove the characteristic disadvantage of microtrons with toroidal cavities, in which the particle energy is always a multiple of U_o.

Let us explain this new property of the microtron in more detail. If $\Omega > 1$, then on the first orbit encircling the cavity, particle velocity is practically equal to the speed of light. In this case, the diameter on the nth orbit is equal to (for m = 2)

$$D_n = (n + 1) \frac{\lambda}{\pi} \qquad (3.10)$$

and depends only on the wavelength of the accelerating field. Consequently the position of the orbits in a microtron along their common diameter, in the first approximation, does not depend on any other parameter except the wavelength. Changing Ω simultaneously and proportionately changes the magnetic field in the accelerator and the energy gain per revolution. With this, it is necessary to proportionately increase the high frequency power exciting the cavity. The position of the orbits does not change, but on each succeeding orbit the energy of the particle increases in proportion to Ω.

Numerical calculations and experiments have shown that in the first type of acceleration one can vary Ω from 0.7 to 1.5, while in the second type of acceleration from 1.8 to 2.2 (for flat cavities with a circular section, m = 2). This means that in the second type of acceleration for a given energy of accelerated particles, the diameter of the accelerator magnet is about half as big as in microtrons of former construction.

SECTION 2 - EQUATIONS OF MOTION

As was shown in Chapter II, particles can be accelerated to the last orbit if their energy and phase on the first orbit corresponded to a region of phase stability. Energy and phase on the first orbit depend on a number of factors; the shape and size of the accelerating cavity, the position of the emitter, the amplitude of the high frequency field, the potential of the constant magnetic field and finally, on the initial phase of the electron (the phase at its exit from the emitter). Thus, resonance acceleration of particles is possible under various combinations of these factors.

This circumstance is useful from the practical standpoint. It is worth noting that when initial attempts to operate in the first and second types of acceleration were made, the accelerator operated normally right from the start, even though the cavity was designed without any detailed calculations.

At the same time, because of their large number, it is practically impossible to find optimum values of the parameter

MICROTRON   47

by cut and dry methods. However, empirical choice of parameters is not necessary since the values can be obtained by means of numerical calculations that describe the initial motion of the electrons in great detail.

Let us investigate the motion of electrons in a cylindrical cavity operating in the E_{010} mode and also a rectangular cavity operating in the E_{011} mode (for these modes, see Ref. 56). The electric field of these modes has only a longitudinal component directed along the axis of the cavity, while the magnetic field has only transverse components (see Fig. 3.2). In practice other cavities of more complicated

(a)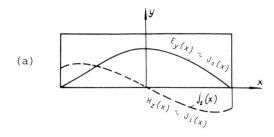

(b)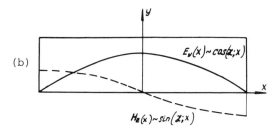

Fig. 3.2 Field distribution in accelerating cavities.
a - Cylindrical cavity; b - Rectangular cavity.

construction can be used. For instance, calculations were performed on acceleration in spheroidal cavities for which it is necessary to calculate the shape of the high frequency field before the equation of motion can be integrated. This complication has no bearing here and at this time we will limit ourselves to calculations of motion in cylindrical and rectangular cavities.

Electric and magnetic fields in cylindrical cavities are described by

$$E_y = E J_0(kR) \cos \omega t \quad , \quad H_x = - E \frac{Z}{R} J_1(kR) \sin \omega t,$$

$$H_z = E \frac{X}{R} J_1(kR) \sin \omega t \quad , \quad E_x = E_z = H_y = 0 \, ,$$

(3.11)

where $R = \sqrt{X^2 + Z^2}$ is the distance from the axis of the cavity (see Fig. 3.2). $J_0(kR)$ and $J_1(kR)$ are Bessel functions of the first kind, $k = 2.405/a$ is the wave number, and a is the radius of the cavity.

In a rectangular cavity the field has a similar form

$$E_y = E \cos k_x X \cos k_z Z \cos \omega t \quad ,$$

$$H_x = - E \frac{k_z}{k} \cos k_x X \sin k_z Z \sin \omega t,$$

$$H_z = E \frac{k_x}{k} \sin k_x X \cos k_z Z \sin \omega t \quad ,$$

$$E_x = E_z = H_y = 0 \quad , \quad (3.12)$$

where $k_x = \pi/a$, $k_z = \pi/b$, a and b are the width and height of the cavity, $k = \sqrt{k_x^2 + k_z^2}$

Substituting these field expressions into the general equation of motion

$$\frac{d}{dt} \frac{mV}{\sqrt{1 - (v^2/c^2)}} = eE + \frac{e}{c} [VH] \quad (3.13)$$

and introducing the dimensionless quantities which were defined in Section 1 of Chapter II, the equation of motion of electrons inside a cavity (in the plane $z = 0$) takes the form

$$\frac{d}{d\phi} \frac{u}{\sqrt{1 - \beta^2}} = - \Omega v + \varepsilon \Omega v F_1(x) \sin \phi \quad , \quad (3.14)$$

$$\frac{d}{d\phi} \frac{v}{\sqrt{1 - \beta^2}} = \Omega u + \varepsilon \Omega F_0(x) \cos \phi - \varepsilon \Omega u F_1(x) \sin \phi \quad .$$

Here we include the following designations: in the cylindrical cavity we assume

$$F_0(x) = J_0(x) \quad , \quad F_1(x) = J_1(x) \quad , \quad (3.15)$$

and in the rectangular cavity

$$F_0(x) = \cos (\kappa x) \quad , \quad F_1(x) = \kappa \sin \kappa x \quad , \quad (3.16)$$

where

$$\kappa = \frac{k_x}{k} = \frac{b}{\sqrt{a^2 + b^2}} \quad .$$

These equations form the basis of all the numerical calculations for the microtron. They were solved for the following initial conditions:

$$x = x_0 \quad , \quad y = 0 \text{ (or } y = \ell), \quad u = v = 0 \quad ,$$

MICROTRON   49

i.e., we considered that the electrons left with zero velocity
the surface of the emitter ($x = x$, $y = 0$ or $y = \ell$) placed in
line with the internal surface of the cavity. Motion inside
the cavity according to Eq. (3.14) proceeds within the limits
$0 < y < \ell$; outside these limits electrons move around the
circumference in the constant magnetic field. This approximation allows us to relate (by simple algebraic equations)
the values of the coordinates and the electron velocity at
the entrance to the cavity on the (n+1) orbit with the values
of the coordinates and the velocity at the exit from the cavity on the nth orbit:

$$u_{n+1} = u_n - \frac{\Omega \ell}{\Gamma_n} ,$$

$$v_{n+1} = - \sqrt{\beta_n^2 - u_{n+1}^2} ,$$

$$x_{n+1} = x_n - \frac{\Gamma_n}{\Omega} (v_n - v_{n+1}) ,$$

$$y_{n+1} = \ell .$$

This motion has a cyclic character and can be easily
programmed for calculation on computers.

The simplifications on which the indicated method of
calculation is based are as follows: field distortion in
the cavity, produced by the thermocathode, the apertures and
holes which connect the cavity to the waveguide are neglected,
and the magnetic field is considered to be uniform. We assume the electron motion to be plane, i.e., no account of
vertical oscillations of the particles is taken. The problem is solved neglecting the interaction of the electrons,
their effect on the field in the cavity and the effect of
electron radiation on the process of acceleration.

SECTION 3 - METHODS OF NUMERICAL CALCULATIONS

The number of orbits in existing microtrons is not large,
usually up to 20-30 orbits. It is thus possible to calculate
motion on all the orbits to the last, verifying the values
of all the parameters. If we consider, however, that the
number of parameters in a microtron on which particle motion
depends is quite large (at least 6), then it would seem that
this method of calculation is not very practical since it
would take a lot of computer time and results would be hard
to interpret. For this reason, from the very start, a number
of methods were developed which substantially reduced the volume of calculations. The essence of these methods depends on
the use of the conditions for resonance accelerations, which
were obtained in Chapter II, and limit the allowable values
of the parameters. Moreover, the results of calculations for
the first several orbits allow one to judge on whether the
particles will reach the last orbit, and whether they will be
accelerated to the corresponding energy.

Complete calculations of motion on all of the orbits is finally performed when an appropriate mode of acceleration has been selected. The aim of this calculation is to determine the optimum size and location of the emitter, to evaluate the tolerances, find the region of initial phase in which particle capture can take place, to determine the limits of energy variation, etc. At this stage, it is reasonable to vary the values of the parameters without imposing any extra limitations.

The number of orbits used for the calculation is determined from the following considerations. In order to preserve accuracy in the case where there is a large number of orbits it is necessary to decrease the integrating step, leading to a rapid increase in computer time. On the other hand, it must be remembered that calculations are based on a number of assumptions. The results of these assumptions may not show up in the first few orbits, but their action is additive and may substantially distort motion on outer orbits. Thus, from our standpoint, it is worthwhile to calculate particle motion for 10-15 orbits, and utilize the theory of phase stability on the remaining ones.

Initial microtron calculations were performed on a trajectograph designed by G. P. Prudkovsky.[31] Thus for the first time it was shown that resonance acceleration of particles in the newly proposed modes of acceleration is possible. From then on, so as to increase accuracy and conserve time, calculations were performed on a general purpose computer. Programming on the EVM (Strela) and later, on the M-20, was performed by E. G. Krutikova. At first, a simplified method of calculation was used where the limits of the parameters were purposely reduced as compared to the maximum possible values. Nevertheless, this method made it possible to calculate orbits from the first and second types of acceleration, which in turn were quickly confirmed by experiment; the method and the results are given in Ref. 4. Later, M. M. Molodensky, with an improved method of calculation, developed the acceleration mode in a strong magnetic field and a spheroidal cavity [the second type of acceleration, $\Omega = 3$ (see also below)].

A more detailed calculation of the first type of acceleration is described in Ref. 14, the results of which are presented below. In this work a number of new accelerating modes were developed, essentially more effective modes which provided a substantial increase of accelerated electrons (see also Ref. 21). In Ref. 22 a more complete method is described which widens the area of the parameters and gives a better approach to the experimentally obtained values.

The main condition for the microtron accelerating mode expressed in the form of Eq. (2.35) is the relationship between parameter ε which is equal to the ratio of the amplitude of the high frequency field to the value of the constant magnetic field [see Eq. (2.7)] and the thickness of the cavity ℓ. It is true, that this relationship also contains values of the transit coordinate x and the equilibrium phase ϕ_s. However, the value of $F_0(x)$ according to Eqs. (3.15) and (3.16) is usually close to unity (since $x \approx 0$) and can be fixed further if the trajectory is known. It is imperative that the value ϕ_s be picked as close to the optimum as possible, where the optimum value is determined by Eq. (2.24) derived from the linear theory of phase oscillations. Numerical calculations show that the average phase of the nonlinear oscillations is somewhat displaced from ϕ_s (see Fig. 2.8). In order to compensate for this shift, the value ϕ_s is selected somewhat larger than that given in Eq. (2.24), namely, we assume

$$\phi_s = 0.35 = 20° \quad . \tag{3.17}$$

As a result, relationship (2.35) takes on the form

$$\varepsilon F_0(x) \sin \frac{\ell}{2} = 0.53 \quad . \tag{3.18}$$

This expression reduces the number of free parameters by one and thus, substantially decreases the amount of calculations.*

Utilizing condition (3.18) we now exclude from investigation all other values of ϕ_s at which resonance acceleration of particles is possible. This limitation is justified, since the region of phase stability is largest (in area) when $\phi_s \approx 20°$ and diminishes as ϕ_s approaches the outer limits in (2.21). Moreover, for the selected value of ϕ_s phase motion is stable against the action of various disturbances. For these reasons, condition (3.17) corresponds to modes of acceleration which are very interesting from the practical standpoint. If we consider that the whole region of equilibrium phase (2.21) is small to begin with, then we can see that the indicated limitation does not appreciably reduce the operating region of the parameters.

Let us now investigate how on the basis of the motion parameters for the first several orbits one can judge whether or not the particles will proceed to the last orbit. In Chapter II it was shown that particles are captured into the resonance mode of acceleration if the points representing them fall inside the region of stability. However, on the first several orbits the size and location on this region depends on the thickness of the cavity and the position of the orbits (see Section 4, Chapter II). Thus, in Ref. 4, it was assumed that points representing the resonance particles in the phase plane $\bar{\phi}$ and γ must lie on a segment determined by

$$0.25 < \bar{\phi} < 0.45 \quad , \quad \gamma = 0 \quad ; \tag{3.19}$$

here $\bar{\phi}$ is the average phase of transit through the cavity, subtracted multiple of 2π, while γ is determined by Eq. (2.13) and the values γ and $\bar{\phi}$ were calculated for the first half-revolution.

---

*Nonlinear theory (see Refs. 96 and 97) makes it possible to choose the value ϕ_s more precisely, but the final results of calculations change insignificantly.

Condition (3.19) considerably reduced the volume of the calculations and allowed us to locate certain accelerating modes which were confirmed by experiment. However, experience has shown that it also greatly reduced the calculated regions of the parameters as compared to the actually existing ones. For this reason we have lately (see Ref. 14) replaced condition (3.19) by another condition

$$0 < \bar{\phi} < 0.5 \;, \quad \gamma = 0 \;, \tag{3.20}$$

in which the values $\bar{\phi}$ and γ were determined on the second passage through the cavity. This condition was used for detailed investigations of the first type of acceleration in a cylindrical cavity. Results of this investigation are given in the following section.

However, for investigating the second type of acceleration the limits set in (3.20) seem to be too stringent. Thus, in Ref. 22, the complete condition for resonance acceleration was used, according to which the graphical point representing resonance particles must lie within the region of phase stability. For the optimum equilibrium phase (3.17) in the plane $\bar{\phi}$ and γ, this region can be considered as being bounded by an ellipse and can be determined by the inequality

$$(\bar{\phi} - 0.25)^2 + 0.84(\bar{\phi} - 0.25)\gamma + 0.48\gamma^2 \leq 0.04 \;. \tag{3.21}$$

The values of $\bar{\phi}$ and γ are usually determined from motion on the second orbit.

We have thus examined the general state of the methods of calculation (a more detailed presentation is given in Refs. 4, 14, and 22). In the following sections, modes of acceleration are described which were calculated by these methods. A data summary of the more interesting modes is presented in Table 3.1, at the end of this Chapter.

SECTION 4 - CALCULATIONS FOR THE FIRST TYPE OF ACCELERATION

In the first report[4] calculation was performed for particles leaving the emitter under initial phases close to the zero corresponding to the maximum accelerating field. More detailed calculations for the first type of acceleration in the cylindrical resonator in which particles with various initial phases were investigated are described in Ref. 14, the results of which are contained in this paragraph.

In calculating the parameters we limited ourselves not only by condition (3.20) but also by the value of x_1 which is the coordinate along which the electrons exit from the cavity. The value of x_1 must be close to zero so that the electrons, on succeeding orbits, will pass through the cavity close to its axis where the electric field changes slowly along the radius and thus, the accelerating process has a high degree of phase stability. On the other hand, far away from the axis of the cavity, an insignificant drift in the orbits could cause changes in the accelerating field and

MICROTRON 53

excite the phase oscillations. For this reason we assume

$$-0.5 \leq x_1 \leq 0.25 \tag{3.22}$$

where, close to the boundary of the interval in condition (3.18), one has to take $F_0(x)$ into account when it is different from unity. The asymmetry in the boundary values of x_1 is due to the circumstance that the curvature of the trajectory causes the points to drift in the direction of $x > x_1$.

In order to simplify the calculations further we place yet another condition: it was assumed that the electrons do not penetrate the cavity through the wall opposite the emitter, i.e., the condition $y_{max} < \ell$ must hold on the first revolution.

Results of the calculations are given in Fig. 3.3 and also in Table 3.1. Operating values of the parameters are presented on the coordinates x_0 and Ω for several fixed values of ϕ_0, but mainly for $\phi_0 = -60°, -30°, 0, 20°$. Initial phases which are not close to zero, for instance, the values for $\phi_0 = -60°$, are also of interest from the practical standpoint, since the use of a hot cathode insures rather great emission in these cases. Values of the cavity thickness ℓ, the exit coordinate x_1 and the average $\bar{\phi}$ are shown on the graph in continuous lines.

Operating regions of the parameters crosshatched on Fig. 3.3 are bounded by the curves $\bar{\phi} = 0$, $x_1 = 0.25$, and $y_{max} = \ell$. The area is largest when $\phi_0 = 0$. On moving to $\phi_0 = -30°$ and $-60°$ the area slowly decreases in size and noticeably shifts in the direction of larger values of x_0; for $\phi_0 > 0$ the area rapidly decreases in size and slightly shifts in the direction of lower values of x_0.

A detailed investigation of these regions leads us to conclude that the operation of a microtron in the first type of acceleration is not critically dependent on the choice of parameters. If the values of the parameters lie within necessary intervals, namely

$$\ell = 1.0-1.2 \quad , \quad x_0 = 1.0-1.6 \quad , \quad \Omega = 0.9-1.3 \quad ,$$

then it is always possible to find a value of the initial phase ϕ_0, for which it is possible to capture particles in some mode of acceleration. In practice this means the following: 1) the thickness of the cavity, the radial dimension of the emitter, and the strength of the magnetic field need not be held to very close tolerances; 2) it is possible to have various deviations from the calculated modes, for instance by biasing the emitter or using a cutoff potential on it; 3) tuning of the accelerator is relatively simple and can be accomplished by changing the high frequency power output, the emitter current, or the magnetic field. The thickness of the cavity can be changed within the limits

$$0.8 < \ell < 1.4 \quad ,$$

but at the ends of this interval it is necessary to choose a corresponding value of x_0. In the early toroidal cavities, the thickness was $\ell = 0.45$–0.65. In the newer cavities, as the thickness ℓ was made larger, there was a significant decrease in the difficulties associated with high frequency breakdown.

The parameter Ω, which for a microtron equipped with a toroidal cavity must be exactly equal to one of the following values, 1, 1/2, 1/3, etc., now can be continuously varied within wide limits. In the first type of acceleration the minimum value of Ω was approximately 0.8; for $\Omega = 0.8$, we have $\Gamma_{s,1} = 2\Omega = 1.6$, i.e., the kinetic energy on the first revolution is equal to 300 keV, and the velocity is 0.8 c. For those small values of energy, Eq. (3.20) is no longer accurate as it becomes necessary to calculate in full not only the second but also the subsequent revolutions since the change in the curvature of the trajectory and the velocity of the electrons could bring about a change in the shape of the phase oscillations (see Section 4, Chapter II). In practice it was possible to obtain a value of $\Omega = 0.65$, which apparently is close to the actual lower limit. In order to obtain values of Ω which are smaller yet, it is necessary to use modes in which $\Gamma_{s,1} = 3\Omega$, 4Ω, etc. If such modes can be discovered, then it would be possible to lower the magnetic field with a consequent reduction of the electron energy at the exit from the accelerator.

Modes of acceleration with lower values of Ω are of interest since they would require a lower amount of power at the cavity, and thus, can be used in a cw microtron. They can also be useful in microtrons operating in the shortwave part of the centimeter waveband (see Appendix I and Ref. 98 for the description of an X-band microtron operating at $\Omega = 0.5$–0.67).

The largest value of Ω is ~ 1.5 and is obtained with the conditions

$$-30° < \phi_0 < 0 \quad , \quad 1.0 < x_0 < 1.4 \quad .$$

For large values of Ω the operating region is bounded by the curves $y_{max} = \ell$. This limitation is not necessarily a matter of principle: on removing it, one can still find modes of acceleration for large values of Ω.

By examining the operating regions, we can see that if the thickness of the cavity ℓ and the position of the cathode x_0 are fixed, then it is possible to vary the magnetic field (proportional to Ω) over a relatively wide range. This is accomplished by varying the power in the accelerating cavity without changing its geometry.

Let us examine this possibility further. The curves $\ell =$ const are slightly inclined to the abscissa. For negative values of ϕ_0 the curves $\ell =$ const shift, along with the operating region, into the direction of larger values of x_0. This results in the curve $\ell =$ const intersecting the straight line $x_0 =$ const for larger values of Ω whereas the points of intersection remain at all times within the operating region. From Fig. 3.3, it follows that a fairly wide range of changing Ω is possible under the conditions of

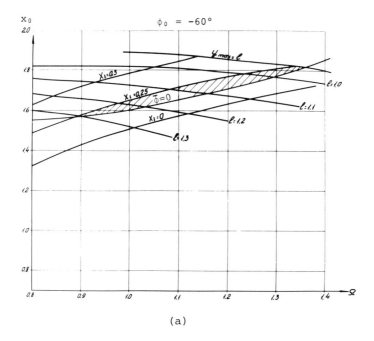

(a)

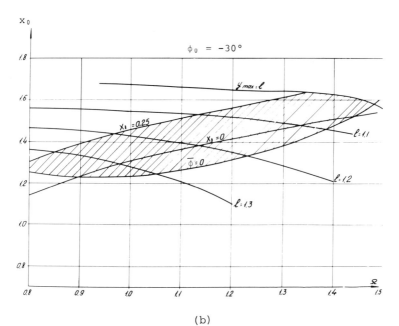

(b)

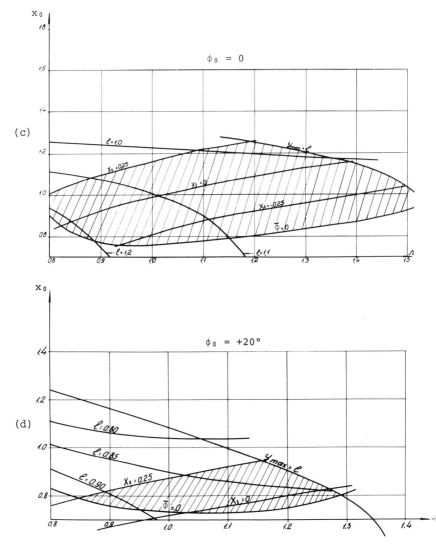

Fig. 3.3 Operating region of the microtron parameters (first type of acceleration).

$$1.0 < x_0 < 1.4 \; ; \; 1.0 < \ell < 1.2 \; .$$

When this is done, as can be seen from Fig. 3.3, the electron exit coordinate x_1 changes and thus the exit aperture should be made in the shape of a horizontal slit. As is evident from the given figures, changing is possible in the range of $0.8 \leq \Omega \leq 1.5$.

MICROTRON   57

In conclusion we would like to point out another notable property of the first type of acceleration which was brought out by the calculations. Figure 3.3 shows that negative values of ϕ_0 results in the curve ℓ = const intersecting the straight lines Ω = const, for larger values of x_0; the points of intersection remain within the operating region. For instance, the curve ℓ = 1.1 intersects curve Ω = 1.1 at the point x_0 = 0.8-1.7, while the exit coordinate varies within the limits - 0.25 < x_1 < 0.25. Consequently it is possible to change not only the magnetic field but also the position of the cathode. Essentially, it means that we can lower the density of the emitted current by using emitters which are elongated horizontally; this will prolong the lifetime, but also will decrease the defocusing action of the space charge. An elongated emitter was proposed in Ref. 14, and was experimentally tried out by K. A. Belovinstev.[90,66]

SECTION 5 - CALCULATIONS FOR THE SECOND TYPE OF ACCELERATION

The geometric properties of the second type of acceleration lead to substantially larger values of Ω, i.e., it allows the use of higher values of magnetic field. At first we limited ourselves to initial phases which were close to zero and the conditions for resonance acceleration were taken in the form (3.19).

On Fig. 3.4 the operating region of the parameters for a cylindrical cavity for ϕ_0 = 0 is shown. This region is

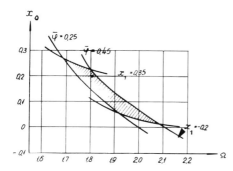

Fig. 3.4 Operating region of the microtron parameters with a cylindrical cavity (second type of acceleration, ϕ_0 = 0).

small as compared to the same region in the first type of acceleration, however, as will be seen below, substitution of condition (3.19) for condition (3.21) causes a substantial increase in the area. Compared to the first type of acceleration, the operating region is shifted in the direction of larger Ω's which is one of the more attractive properties of these modes. In particular, a detailed calculation of the mode (on Fig. 3.4 it is designated by a dot) in which Ω = 1.8, i.e., the energy of the accelerated electrons is 1.8 times greater than in a microtron with a toroidal cavity and a magnet of similar dimensions. Within the operating region, there are points which correspond to a value of $\Omega = 2$.

By changing the shape of the cavity it is possible to further increase the magnetic field. Thus, in Ref. 4, we have calculated the motion in a rectangular cavity which is elongated in the horizontal plane. [In Eq. (3.16), $\kappa = 0.55$.] The operating region of the parameters is shown in Fig. 3.5;

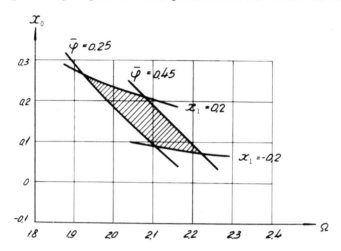

Fig. 3.5 Operating region of the microtron parameters with a rectangular cavity (second type of acceleration, $\phi_0 = 0$).

it is evident that the region is displaced even further in the direction of larger values of Ω. On Fig. 3.6 electron trajectories in this cavity are shown; in Fig. 3.7 results

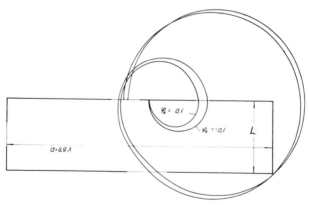

Fig. 3.6 Electron trajectories in a rectangular cavity.

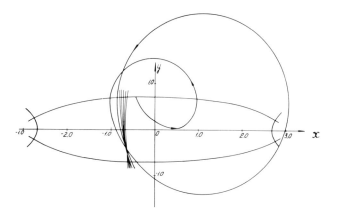

Fig. 3.7 Electron trajectories in a spheroidal cavity ($\Omega = 3$, $\ell = 1.4$, $\varepsilon = 0.9$).

obtained by M. M. Molodensky are given for a spheroidal cavity, the walls of which are sections of ellipses and hyperbolae of revolution. In such a cavity, calculations show that it is possible to have acceleration for $\Omega = 3$.

A more complete calculation of the second type of acceleration utilizing condition (3.21) is given in Ref. 22. By means of substituting condition (3.19) for condition (3.21) the parameter limits can be considerably increased.

As an example, on Fig. 3.8 operating regions of the parameters calculated by both methods for a rectangular resonator are shown. The crosshatched area represents condition (3.19). Lines Φ = const are found through the use of condition (3.21), where Φ is the width of the interval of the initial phase of

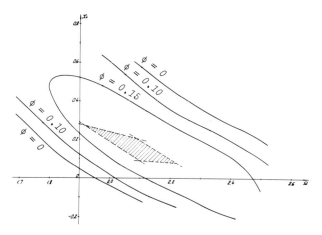

Fig. 3.8 Operating regions of the microtron parameters with a rectangular cavity (second type of acceleration).

the electron (e.g., the phase at the exit from the emitter), for which the parameters of motion satisfy this condition. With the aid of the methods indicated a series of new modes with strong magnetic fields were calculated.[22] Calculations have shown that large values of Ω correspond to modes in which the emitter is placed on the left side of the axis of the cavity, or the cavity is elongated in the horizontal plane. On Fig. 3.9 the trajectory of particles in a circular cavity

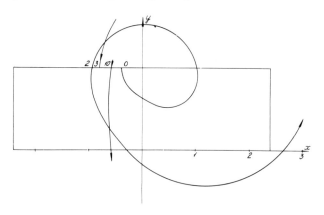

Fig. 3.9 Motion of particles in a cylindrical cavity, emergent apart from the cavity axis ($\Omega = 3$, $\ell = 1.5$, $\varepsilon = 0.85$, $x_0 = 0.40$, $\phi_0 = 3.00$).

in which the cathode is displaced to the left of its axis is shown for mode $\Omega = 3$. The disadvantage of this mode is that the electron orbits pass at a distance from the axis of the cavity where the electric field changes rapidly with radius, and as a result, a slight radial drift of the orbits produced by magnetic field nonhomogeneity can destroy the phase motion. The mode of acceleration in a spherical cavity (Fig. 3.7) also has the same disadvantage.

By elongating the cavity horizontally it is possible to increase the parameter Ω to 3 without displacing the emitter. The electron trajectory in such a cavity intersects the cavity corner (see Fig. 3.10), making it necessary to deform the wall of the cavity at that point. It would seem that this mode would be easier to obtain in practice than those with a displaced emitter. In such an elongated cavity, if one were to displace the cathode to the left of its axis, then the particles would not be able to round the cavity and would instead pass through it (Fig. 3.11). Calculations show that such a supplementary pass makes it possible to increase the parameter Ω. Indeed, calculations for modes with the parameters $\Omega = 4$ and 5 were made. For these modes, however, particles pass close to the cathode and this raises practical difficulties.

MICROTRON   61

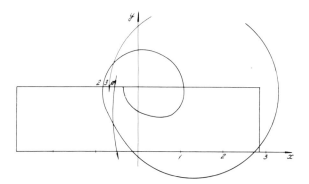

Fig. 3.10 Particle injection in a rectangular cavity elongated in the horizontal direction ($\Omega = 3$, $\ell = 1.47$, $\kappa = 0.55$, $\varepsilon = 0.82$, $x_0 = 0.30$, $\phi_0 = 3.00$ and 2.95).

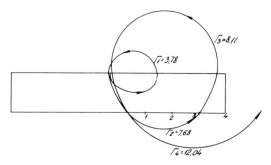

Fig. 3.11 Injection in a rectangular cavity with a supplementary orifice for the particles near the end of the cavity ($\Omega = 4$, $\ell = 1.435$, $\kappa = 0.39$, $x_0 = 0.30$, $\phi_0 = 2.65$).

SECTION 6 - MICROTRON EFFICIENCY AND MEANS OF INCREASING IT

The efficiency of a microtron is determined by the number of emitted electrons which reach the last orbit. This number in turn is determined by two factors: the size of the region of capture and the vertical aperture. The region of capture is that region of initial phases of the electrons captured in a resonance mode of acceleration; we will call them resonance particles. Since in practice there is a loss of not only nonresonance electrons, but also those resonance electrons which have large vertical amplitudes, then the number of accelerated electrons is determined by the ratio between the allowable vertical dimension of the emitter and its true size.

Vertical losses will be discussed at the end of this section, but first, let us determine the size of the capture region. The region of capture can be approximately evaluated by means of (3.21), but a more accurate determination is obtained by integrating the equation of particle motion all the way to the last orbit. Similar calculations[4] have shown that the region of capture consists of approximately 4° in the first type, and 13° in the second type of acceleration. The results of the calculations for the first type of acceleration are shown on Fig. 2.8. The narrow width of the region of capture decreases the accelerated current and the efficiency of the accelerator, and in order to increase the current it is necessary to increase the emission of the cathode.

As has already been pointed out above, the first calculated modes of acceleration[4] corresponded to a zero initial phase. Subsequent calculations[14,21] have shown that negative initial phases substantially widen the region of capture in the first type of acceleration. For example, in the acceleration mode where $\ell = 1.07$, $x_0 = 1.7$, and $\Omega = 1.2$ (a cylindrical cavity), the initial phases of resonance electrons are contained within the limits $-75° < \phi_0 < -50°$, i.e., the width of the region of capture is about six times larger than for $\phi_0 = 0$. The reason for such an increase is that the width of the region of capture, according to the calculations, depends on the allowable energy spread $\Delta\Gamma$ (on the first orbit), determined by the size of region (3.21) and is approximately $\pm 0.05 \Omega$. Figure 3.12 shows a schematic representation of

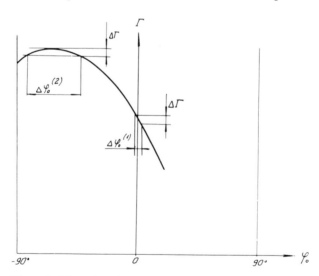

Fig. 3.12 The dependence of the electron energy after the first orbit on the initial phase.

the energy of the particles on the first exit from the cavity as functions of the initial phase. It is seen that for negative values of ϕ_0 on the curve $\Gamma(\phi_0)$ there exists a maximum point, and by making use of this maximum point it is possible to substantially widen the region of capture. The area of accelerator parameters corresponding to negative values of

MICROTRON   63

ϕ_0 is shown on Fig. 3.3. This area is narrower than the one
for the zero initial phase (Fig. 3.3) and for this reason the
emitter must be placed further from the axis of the cavity.
    Because of the Schottky effect, emission density is
greatest for zero phase and decreases as ϕ_0 becomes negative
(Fig. 3.13). However, this decrease is not very significant

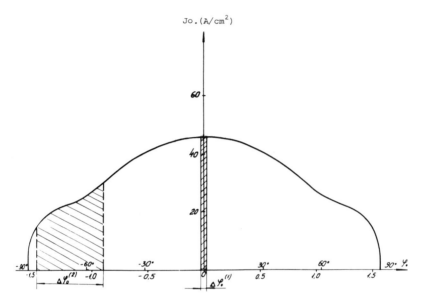

Fig. 3.13 The dependence of emission on the initial
phase (T = 1600°C, E (kV/cm) = 150 cos ϕ_0).

and in the final analysis the increase in the region of capture produces a substantial gain in the number of resonance
electrons. This calculated result was verified experimentally
on our first microtron as well as on the 17-orbit microtron
(see Chapter VII).
    A mode of acceleration with similar characteristics was
determined[22] for the second type of acceleration. In this
case after the first half-revolution, the curve $\Gamma = \Gamma(\phi_0)$
also has a maximum point for negative value of ϕ_0. After the
ensuing pass (the first orbit completely encircling the cavity) the representing points on the phase plane which correspond to this maximum point form a characteristic bump (Fig.
3.14). From this figure it can be seen that for the mode of
acceleration which has the parameters $\ell = 1.65$, $x_0 = 0.4$,
$\Omega = 1.8$, this bump is located within the ellipse (3.21), and
the width of the region of capture is greatest, encompassing
approximately 50°, i.e., four times larger than the region
for a zero initial phase.
    It is worthwhile to point out that modes which have nega-
tive initial phases are useful from another standpoint. As
we have seen, the region of capture corresponds to the maximum

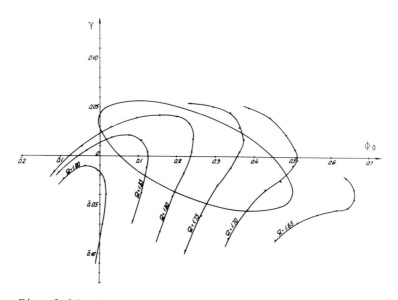

Fig. 3.14 The curves $\Gamma = \Gamma(\phi_0)$ in the phase plane (second type of acceleration).

point on the curve $\Gamma(\phi_0)$. Thus, on the first half-revolution, resonance electrons gain the largest possible amount of energy leading to a higher efficiency of the accelerator.

Let us now consider the vertical motion of the particles. Particles proceeding on the first and subsequent orbits are usually lost if the amplitude of their vertical oscillations exceeds the maximum allowable (an exceptional case is when there is a significant radial drift due to a nonhomogeneity of the field which produces radial losses). There are two types of vertical losses. The first is that all nonresonance particles are lost because of the excitation of phase oscillations, these particles came to the wrong phase of the high frequency field and are defocused in the vertical direction and are lost on the walls of the cavity. The second type occurs if the vertical dimension of the emitter is too large. Then the resonance electrons which emerge from its vertical edges are lost since the amplitude of their oscillations is larger than the vertical aperture.

The first type of losses depends on the size of the region of capture which we determined above. Losses of the second type depend not only on the vertical size of the emitter, but also on the condition of the motion on the first half of the revolution inside the cavity. The point is that an electron which is not moving along the median plane experiences the action of the vertical force which is due to the horizontal component of the alternating magnetic field. Linearized equations of vertical motion inside a cavity have the form

$$\frac{dz}{d\phi} = \frac{p}{\Gamma} \quad , \quad \frac{dp}{d\phi} = Mz \quad , \tag{3.23}$$

where

$M = \varepsilon\Omega v \dfrac{J_1(x)}{x} \sin \phi$ in the circular cavity,

$M = \varepsilon\Omega v (1 - \kappa^2) \cos \kappa x \cdot \sin \phi$ in the rectangular cavity.

In the program for making the calculations, the possibility was foreseen for numerically integrating Eq. (3.23) simultaneously with integrating the equation of undisturbed motion, and the value M was determined by this undisturbed motion.

Thus we obtained the transformation matrix of the vertical coordinates and the momentum of electrons from the emitter to the exit of the cavity. Let us examine for example the first type of acceleration where $x_0 = 1.2$, $\ell = 1.0$, $\Omega = 1.1$, and $\phi_0 = 0$; calculations lead to the matrix transform:

$$\begin{pmatrix} z_1 \\ p_1 \end{pmatrix} = \begin{pmatrix} 1.83 & 4.03 \\ 0.694 & 2.08 \end{pmatrix} \begin{pmatrix} z_0 \\ p_0 \end{pmatrix} \tag{3.24}$$

which shows that the alternating magnetic field produces a noticeable defocusing effect. Indeed, if we take the entrance aperture used, $-0.25 < z < 0.25$, then the vertical momentum on the exit from the cavity must satisfy the condition $|p_1| < 0.05$, otherwise the beam is lost at the entrance aperture on the very first revolution. If the exit aperture is made in the shape of a horizontal slit then the allowable size of the emitter, taking into account the defocusing action of the slit (see Chapter IV), is determined by the inequality $|z_0| \leq 0.025$, i.e., approximately 1.7 times less than the size of the emitter used in practice (see, for example, Ref. 3). Because of this, 40% of the beam must be lost on the first revolution. In practice, even larger particle losses are observed. These losses are associated with the effect of space charge and distortion of the electric field close to the edges of the emitter. The situation can be improved if a vertical slit is used for the exit aperture. In that case, the aperture does not introduce additional defocusing and, from Eq. (3.24), it follows that the inequality becomes $|z_0| \leq 0.045$, which happens to coincide with the actual size of the emitter.

By decreasing the vertical dimension of the emitter, one can also decrease the losses but only within certain limits. If this dimension becomes comparable to the gap between the emitter and the lid, then the electric field in the vicinity of the emitter is noticeably distorted, and the value of p_0 in Eq. (3.24) increases and defocusing again becomes stronger. Defocusing can be substantially reduced by placing the cathode deeper inside the wall; in this case an electric lens will be formed in the vicinity of the emitter which will effectively focus the electron beam. Such a method was initially used on our first microtron (see Ref. 16) and has been used on all subsequent microtrons (see, for example, Ref. 66).

From all that was said above, it follows that there are certain rigid requirements in positioning the emitter and the cavity. The median plane of the cavity must be exactly perpendicular to the magnetic field. The center of the emitter must be placed on the median line and the surface of the emitter must be parallel to the internal surface of the cavity lid. The latter requirements is quite essential; it can be, however, somewhat loosened if the emitter is placed deeper within the wall.

The allowable dimensions of the emitter are comparatively small. As a result, the electron bunches have small dimensions and a high density, considerably higher than in linear accelerators. However, this also means that the emitter in a microtron operates under conditions of high density of current, a density that is crucial in determining the lifetime of the emitter.

TABLE 3.1 Parameters of Calculated Accelerating Mode

No.	Type of Acceleration	Cavity	Ω	ε	x_0	ℓ	P_r, kW (for λ = 10 cm)	Φ
1	1	Cylindrical	0.80	0.963	1.0	1.17	100	0.09
2	1	Cylindrical	1.10	1.06	1.10	1.05	260	0.07
3	1	Cylindrical	1.10	1.11	1.20	1.00	300	0.07
4	1	Cylindrical	0.7-1.5	0.94	1.0	1.2	Modes with varying energies	
5	1	Cylindrical	0.9-1.7	0.94	1.15	1.2		
6	1	Cylindrical	1.2	1.04	1.70	1.07	250	0.4
7	1	Rectangular $\kappa = 0.9$	1.0	0.86	1.10	1.4	180	0.7
8	2	Cylindrical	1.8	0.80	0.20	1.45	450	0.24
9	2	Rectangular $\kappa = 0.55$	2.0	0.79	0.20	1.47	550	0.21
10	2	Spheroidal	3.0	0.90	-0.43	1.4	1500	0.25
11	2	Cylindrical	3.0	0.85	-0.40	1.50	1300	0.1
12	2	Rectangular $\kappa = 0.55$	3.0	0.82	-0.30	1.47	1400	0.1
13	2	Rectangular $\kappa = 0.40$	3.0	0.78	0.20	1.50	1700	0.1
14	2	Rectangular	4.0	0.81	-0.30	1.43	3400	0.3

CHAPTER IV

PARTICLE FOCUSING

SECTION 1 - VERTICAL AND RADIAL MOTION

In contrast with the majority of other cyclic accelerators, particles in the microtron are not focused by the magnetic field, but rather by the electric field in the cavity. For this reason, the properties of the transverse motion in a microtron are noticeably different from the properties of betatron oscillations in conventional accelerators. Vertical and radial motion in the microtron was studied both theoretically and experimentally by one of the authors of this book (see Refs. 14 and 5) and in this Chapter these results are presented.

Let us first consider vertical focusing. In a microtron, the magnetic field is uniform and it is well known that vertical motion in such a field is unstable; since all the particles have a vertical component of velocity, they move along helical lines and are lost. Thus vertical stability in a microtron is completely determined by the focusing action of the accelerating cavity.

In Fig. 4.1 a section of an accelerating cavity with the lines of force of the electric field is shown. It is evident that at the entrance to the cavity electrons are focused while at the exit they are defocused by the transverse component of the electric field. The equilibrium phase in a microtron is positive, i.e., it corresponds to a decreasing high frequency electric field. For this reason, relativistic electrons are noticeably defocused inside the cavity by the alternating magnetic field, and secondly, the focusing field which acts on the particles close to the entrance aperture is stronger than the defocusing field acting on them close to the exit aperture.

The magnitude of the focusing and defocusing forces depends on the shape of the apertures and the shape of the cavity. In early microtrons, axisymmetric cavities were used (toroidal and cylindrical), with circular holes which insured focusing of particles in the vertical and the radial directions. However, this focusing is rather weak, and at high energies soon becomes ineffective. As a result, when the particles approach the speed of light, the focusing by the transverse electric field (close to the entrance aperture) is more strictly compensated by the defocusing caused by the alternating magnetic field inside the cavity and the alternating electric field close to the exit aperture.

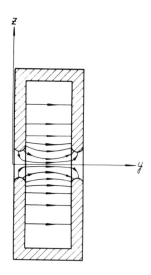

Fig. 4.1 Electric field flux lines in an accelerating cavity.

For comparison, let us consider the properties of transverse motion in a linear electron accelerator. The alternating field in the cell of an irised waveguide of a linac has the same structure as the field in a cylindrical cavity of a microtron. However, in a linac the equilibrium phase corresponds to a field rising in time which defocuses the particles. At higher particle energies this defocusing becomes weaker for the same reason that focusing in a microtron becomes weaker. Because of this weaker defocusing the transverse momentum of the particles ceases to grow; at the same time their longitudinal momentum continues to grow under the action of the accelerating field. This results in a smaller angle between the particle trajectory and the axis of the linac (the tangent of this angle is equal to the ratio between the transverse momentum and the longitudinal momentum). In the final analysis the radial deviation of the particles increases only logarithmically with gain in energy, i.e., very slowly.

In the microtron the situation is different. Although the high frequency field focuses the particles, nevertheless, the vertical dimension of the beam increases much faster than in a linac. This is due to the fact that the orbit length in a microtron increases in proportion to the particle energy. If the vertical momentum were to remain constant, then the angle between the trajectory and the horizontal plane would decrease inversely proportional to the longitudinal momentum. In this case, as the orbits become longer, the particles would move along helical lines of increasing radius. Actually, as we will see below, the vertical motion is of an oscillating nature due to the focusing action of the cavity.

Vertical focusing in a microtron with a circular cavity and circular aperture was first calculated by Bell.[61] Bell showed that it is imperative to take into account the small variation in the vertical coordinate inside the cavity resulting from the focusing action of the electric field on the particle as it enters into the cavity. If this variation were neglected then the change of the particle per revolution would be inversely proportional to the square of the energy. This would mean that the particles are not focused at all, and the beam would gradually spread out as was shown above. Because in the actual case the particles leave the cavity closer to the axis than when they enter it, this defocusing action decreases; the change of the vertical momentum per revolution decreases inversely proportional to the energy. Finally, vertical oscillations are excited, which, because of the weakening of the focusing, cause the oscillation frequency to decrease inversely proportional to the square root of the energy while their amplitude grows in proportion to the fourth root of the energy. In this way particle focusing in a microtron is such that as the energy increases the vertical beam size grows faster than in a linac. This growth causes particles to be lost. At the same time, there are other sources of losses (cavity misalignment, nonhomogeneity of the magnetic field, etc.), which are particularly significant in microtrons with large numbers of orbits.

Thus, for instance, in the 56-orbit microtron,[60] it is shown that even slight misalignment (the order of 15 arc seconds) of the axis of the cavity to the median plane of the magnet causes a substantial loss of electrons when they are accelerated to 29 MeV. In order to insure vertical stability in this microtron it was necessary to create a local inhomogeneity in the magnetic field close to the cavity in spite of the fact that this caused particle losses due to the phase oscillation excitation. The resulting beam intensity in this accelerator turned out to be very low (0.05 mA).

In order to improve focusing it is possible to use a split sector magnet introducing edge focusing.[62] This was done by Brannen and Fröelich[63] who constructed a split microtron of 7 orbits. However, it is not necessary to use a split magnet in order to improve vertical focusing. One of the authors of this book has shown[5] that vertical focusing could be made stronger by getting away from axial symmetry in the cavity and by using slit-shaped flight apertures instead of circular ones.

Indeed, the former considerations on the weakening of vertical focusing with the increase in particle energy are only true for the axially symmetrical cavities. But the situation is quite different for other shaped cavities or flight apertures. Essentially, the vertical component of the electric field in the vicinity of a horizontal slit is twice as large as in a circular aperture, while in the case of a vertical slit it is absent altogether (see Fig. 4.2). By combining, for instance, an entrance horizontal slit with an exit vertical slit, it is possible to have powerful vertical focusing. Indeed in this case the focusing action of the cavity can be too great, thus resulting in an onset of

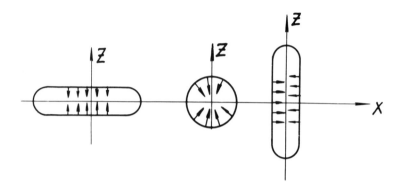

Fig. 4.2 The electric field close to the flight aperture.

vertical instability due to overfocusing.

Calculations presented in the following sections will show the shape of apertures which give optimum focusing. For an optimized shape aperture the change in the vertical momentum which is determined by the action of the alternating field in the cavity weakly depends on the particle energy and, in practice, after the first revolutions, particle focusing does not decrease. On further acceleration the period of the vertical oscillations expressed in units of orbits remains unchanged, and in this respect the microtron is similar to other cyclic accelerators.

The increase in the orbit length in a microtron leads to the fact that the amplitude of the vertical oscillations and the vertical momentum remain unchanged, i.e., there is no adiabatic damping. At the same time, the vertical angle oscillates with amplitude which decreases in inverse proportion to the energy.

Let us now pass to the radial motion. As we know, a uniform magnetic field produces radial focusing: in such a field the particles return to the initial point independently of the exit angle. However, motion in such a field possesses a $n = 0$ resonance instability where any small perturbation in the field substantially disturbs motion if the number of revolutions is large enough. This property of the motion can be easily understood if one considers that in a uniform field, the position of the orbit is not at all fixed and for this reason there is no restoring force.

The accelerating cavity in the first approximation can be treated as a localized force. It is easy to see that in a uniform field the orbit will rotate relative to the point where this force is applied and will asymptotically approach such a position in which the momentum increase is tangent to the circumference (Fig. 4.3).

Consider, for example, a cavity with flight apertures in the shape of horizontal slits. In such a cavity, the particles at each revolution are focused in the vertical direction by the high frequency field and defocused in the radial

MICROTRON 71

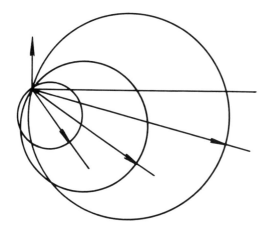

Fig. 4.3 Orbit displacement in a uniform magnetic field under the action of an electric field.

direction. The corresponding increase in the transverse momentum grows almost linearly with distance from the axis of the cavity, and after several revolutions becomes practically independent of the particle energy. The increase in the longitudinal momentum is weakly dependent on the energy and is almost constant close to the axis of the cavity.

From this we can conclude that the orbits near the cavity will be situated as shown schematically in Fig. 4.4. The trajectory of each particle asymptotically approaches the equilibrium position for which the tangent of the angle of inclination of the trajectory with respect to the axis of the cavity is equal to the ratio of the transverse and longitudinal

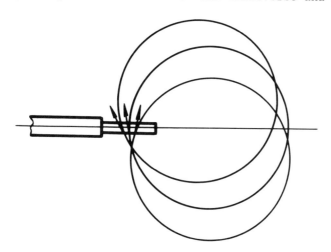

Fig. 4.4 Radial displacement of the orbit.

momentum increments. In this way the total momentum increase is directed along the equilibrium orbits.

We should mention that such orbit inclination does not appreciably affect the operation of the microtron. One only needs to watch that the particles on their last orbit do not strike the side of the vacuum chamber. Aside from this the inclination of the trajectory should not change the energy gain which would destroy phase stability. Both of these limitations are weak and in practice the last orbit can be displaced perpendicularly to the common diameter by several centimeters.

Much more significant is orbit displacement along their common diameter. Such orbit drift is caused by asymmetry in the magnetic field relative to that direction. With this kind of a displacement there is a change in the energy gain per revolution (due to the drop-off in the accelerating field along the radius), and the particles can leave the phase stable region. The limiting drift depends on the shape of the cavity and the flight apertures and usually amounts to several millimeters. It is easy to see that the cavity cannot, in any real sense, diminish this drift. This becomes clear from the following considerations: if one neglects the thickness of the cavity and replaces it by an accelerating gap of zero thickness, then the particles return to the original point independently of the exit angle from the cavity which is determined by the alternating field in it.

In Section 3 of this Chapter, a calculation is made for the radial motion in arbitrarily shaped flight apertures and a finite thickness accelerating cavity. These calculations show that in the actual case the orbits are asymptotically attracted to the axis of the cavity and that the radial dimensions of the beam decrease. However, this effect is of the order $1/n^2$ (n is the number of orbits), i.e., it quickly diminishes as the particle energy increases. Thus, it would appear that the requirements for field uniformity as regards radial drift along the common diameter of the orbits are quite strong in the microtron.

By changing the shape of the flight apertures in order to strengthen vertical focusing we simultaneously increase radial defocusing. Calculations presented below indicate that when this happens there is an increase in the radial angular spread in the accelerated beam, however radial stability is conserved. Magnetic field uniformity requirements for variously shaped flight apertures remain essentially the same.

SECTION 2 - CALCULATIONS FOR VERTICAL FOCUSING

Calculations for vertical focusing are performed by the matrix method presented in Appendix II which makes use of the fact that each complete revolution of an electron is a single cycle of the transverse motion which is described by two second-order matrices (in the linear approximation investigated by us, vertical and radial motion are independent of each other since the cavity is symmetric with respect to the vertical and horizontal diametric planes). For a constant energy the matrix elements would be constant, while the energy increase of the particle leads to the fact that the matrix coefficients are slow monotonic functions of the number of orbits. Each matrix determines the property of the corresponding motion (see Section 1, Chapter II).

The transformation matrix for the coordinates and the momentum for a complete revolution is a product of two matrices describing motion through the cavity and outside it. The first matrix has essentially the same form for vertical and radial motion; the second, however, differs substantially for these motions. Indeed, the change in the vertical coordinate for motion outside the cavity is proportional to the momentum of the particle at the exit from the cavity, while the change in the radial coordinate is relatively small, and in the first approximation, the particle returns to its point of origin independently of the value of the radial momentum. This points out the qualitative difference in the character of the two motions. Calculations show that vertical motion has an oscillatory nature while radial motion is nearly monotonic.

In analyzing motion through the cavity, we assume that its thickness is substantially larger than the minimal dimensions of each of the flight apertures. Thus, the length of the nonuniform electric field close to the apertures is small in comparison with the thickness of the cavity, and for this reason each of the apertures can be approximately treated as a thin electrostatic lens in which the field corresponds to the alternating electric field at the moment the electron passes through the plane of the aperture,* while the field inside the cavity may be considered undisturbed. In contrast to Bell,[61] we obtain the matrix for the cavity and the apertures not directly, but as a result of products of matrices for three regions of motion, the entrance aperture, the undisturbed cavity, and the exit aperture. Our approach allows us to investigate focusing in the general case where the cavity does not necessarily have any symmetry of revolution.

We will solve the problem at hand by making use of the power series expansion with a small parameter $\Omega/\Gamma \approx 1/n$, where n is the number of orbits and Ω and Γ are the dimensionless magnetic field and the total particle energy respectively [see Eqs. (2.5) and (2.8)]. The results obtained are useful only after several revolutions when the particles become relativistic ($\Gamma \gg 1$). Here again, as in Chapter III, we do not consider any radiation effects.

The equation of vertical particle motion is:

$$\frac{d}{dt} \frac{mV_z}{\sqrt{1 - \beta^2}} = eE_z + \frac{e}{c}(V_x H_y - V_y H_x) \quad . \tag{4.1}$$

---

*In calculating the passage of the particles through the aperture it would have been possible to introduce the transit factor into Eq. (2.37), however it can always be replaced by unity since the smallest dimension of the aperture is of the order $\lambda/15$ (see Ref. 14).

The corresponding system of coordinates is shown on Fig. 4.5 where a section of the accelerating cavity along the median plane is given.

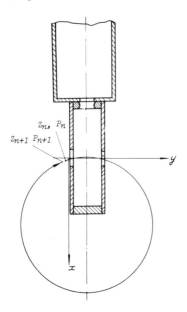

Fig. 4.5 For calculating vertical motion.

Let us first calculate focusing of the electrons by the undistrubed magnetic field in the cavity. Close to the median plane, the components of the high frequency field are determined by the following expressions

$$E_y = E F_0(x) \cos \phi \quad , \quad H_z = E F_1(x) \sin \phi ,$$
$$H_x = - E F_2(x) z \sin \phi \quad , \quad E_x = E_z = H_y = 0 . \quad (4.2)$$

Here x, z, and ϕ are the dimensionless coordinates and time introduced in Section 1 of Chapter II, E is the amplitude of the field, the functions F_0, F_1, and F_2 which determine the space distribution of the field component depend on the shape of the cavity, and they are related by the equation

$$F_2(x) = F_0(x) - \frac{dF_1}{dx} , \qquad (4.3)$$

which directly follows from Maxwell's equation.

In dimensionless variables, Eq. (4.1) for motion inside the cavity takes the following form:

$$\frac{dz}{d\phi} = \frac{p}{\Gamma} \quad , \quad \frac{dp}{d\phi} = \epsilon \Omega v \, F_2(x) \, z \sin \phi . \qquad (4.4)$$

MICROTRON   75

We will designate the values at the entrance to the cavity
by a single prime (Γ', ϕ', ...), while at the exit from the
cavity we will designate them by double primes (Γ'', ϕ'', ...).
Let us introduce the small parameter

$$\mu = \Omega/\Gamma' \tag{4.5}$$

and look for a solution to the system of Eqs. (4.4) in the
form of a series:

$$z = z^{(0)} + \mu z^{(1)} + \mu^2 z^{(2)} + \ldots ,$$
$$p = p^{(0)} + \mu p^{(1)} + \mu^2 p^{(2)} + \ldots . \tag{4.6}$$

The members of these series must satisfy the initial conditions

$$\begin{aligned} z^{(0)} = z', \quad z^{(1)} = z^{(2)} = \ldots = 0 \\ p^{(0)} = p', \quad p^{(1)} = p^{(2)} = \ldots = 0 \end{aligned} \quad \text{at } \phi = \phi',$$

where $z^{(0)}$ and $p^{(0)}$ are values of one order and do not depend
on the energy. According to Eq. (2.10) we have

$$\Delta z = \frac{p}{\Gamma} \tau = \frac{2\pi}{\Omega} p \sim p .$$

We consider the vertical oscillations to be small and
limit ourselves to a linear approximation; in this way one
can find the dependence $\Gamma(\phi)$ from the undisturbed motion along
the median plane and also make $v = 1$. From Eq. (3.14) we get

$$\frac{d\Gamma}{d\phi} = \varepsilon\Omega \cdot F_0(x) \cos \phi$$

from which

$$\Gamma(\phi) = \Gamma' [1 + \mu\varepsilon \cdot F_0(x) (\sin \phi - \sin \phi')] . \tag{4.7}$$

Substituting expressions (4.6) and (4.7) into the system of
Eqs. (4.4), equating the coefficients of the same order elements of μ, and then integrating the differential equation,
we subsequently get the functions $z^{(0)}$, $p^{(0)}$, $z^{(1)}$, $p^{(1)}$, etc.
In our case it is adequate to limit ourselves to the elements
$z^{(1)}$ and $p^{(1)}$. By making use of the solutions we find the
matrix (β_{ik}) which relates the vertical coordinate and the
momentum before and after passage through the undisturbed
field in the cavity:

$$\begin{pmatrix} z'' \\ p'' \end{pmatrix} = \begin{pmatrix} \beta_{11} & \beta_{12} \\ \beta_{21} & \beta_{22} \end{pmatrix} \begin{pmatrix} z' \\ p' \end{pmatrix}$$

The elements of β_{ik} are determined by the equations

$$\beta_{11} = 1 + \mu\varepsilon \cdot F_2(x)(\ell\cos\phi' - \sin\phi'' + \sin\phi') ,$$

$$\beta_{12} = \mu\frac{\ell}{\Omega} ,$$

$$\beta_{21} = \varepsilon\Omega\, F_2(x)(\cos\phi' - \cos\phi'') + \qquad(4.8)$$

$$+ \mu\varepsilon^2\Omega\, F_2^2(x)\left[\int_{\phi'}^{\phi''}\cos^2\phi\, d\phi + \sin\ell - \ell(1+\cos\phi'\cos\phi'')\right] ,$$

$$\beta_{22} = 1 + \mu\varepsilon \cdot F_2(x)(\sin\phi'' - \sin\phi' - \ell\cos\phi'') ,$$

where $\ell = \phi'' - \phi'$, the dimensionless thickness of the cavity.

Let us now consider the action of the beam apertures. The smallest transverse dimension of the apertures is small as compared to the wavelength. The dimension of the nonuniform region of the field which contains the vertical component E_z is approximately equal to this dimension of the aperture. Let us treat the apertures as thin electrostatic lenses. We assume that their field satisfies the equation (see Section 3)

$$\frac{\partial E_z}{\partial z} = -\alpha\frac{\partial E_y}{\partial y} , \qquad(4.9)$$

where α is a constant coefficient which depends on the shape of the apertures. In this case, the field close to the median plane is

$$E_z = -\alpha z\frac{\partial E_y}{\partial y} .$$

Let us integrate Eq. (4.1) by neglecting the action of the magnetic field and keeping in mind that the lens is thin, i.e. neglecting the change in the vertical coordinate during passage through the lens; in dimensionless variables we obtain

$$\Delta p = \mp\frac{\alpha\varepsilon\Omega}{v}F_0(x)\cos\phi . \qquad(4.10)$$

Assuming that $v = 1$, approximately, let us now write the transformation matrices for the coordinates and the momentum at the entrance and exit apertures

$$\begin{pmatrix}z^*\\p^*\end{pmatrix} = \begin{pmatrix}1 & 0\\ -\varepsilon\Omega\alpha'F_0\cos\phi' & 1\end{pmatrix}\begin{pmatrix}z\\p\end{pmatrix} \qquad(4.11)$$

$$\begin{pmatrix} z^* \\ p^* \end{pmatrix} = \begin{pmatrix} 1 & 0 \\ \varepsilon\Omega\alpha"F_0 \cos \phi" & 1 \end{pmatrix} \begin{pmatrix} z \\ p \end{pmatrix} \qquad (4.11)$$

where z and p are the coordinates and the momentum before the aperture and z^* and p^* are the coordinates and the momentum after the aperture; α' and $\alpha"$ are the α coefficients at the entrance and the exit apertures respectively.

For electron motion in the constant magnetic field outside the cavity

$$p = \text{const} , \quad z^* = z + \frac{p}{\Gamma}\left(\frac{2\pi\Gamma}{\Omega} - \ell\right) ,$$

thus, the transformation matrix is equal to

$$\begin{pmatrix} 1 & \frac{2\pi - \mu\ell}{\Omega} \\ 0 & 1 \end{pmatrix} . \qquad (4.12)$$

By multiplying the four matrices (4.8), (4.11), and (4.12), we obtain the matrix α_{ik} describing a complete revolution of the electron in the microtron. By choosing the beginning and the end of the revolution as shown in Fig. 4.5, we get

$$\begin{pmatrix} z_{n+1} \\ p_{n+1} \end{pmatrix} = \begin{pmatrix} \alpha_{11} & \alpha_{12} \\ \alpha_{21} & \alpha_{22} \end{pmatrix} \begin{pmatrix} z_n \\ p_n \end{pmatrix}$$

where

$$\alpha_{11} = 2 + \frac{2\pi}{\Omega}\alpha_{21} - \alpha_{22} ,$$

$$\alpha_{12} = \frac{2\pi}{\Omega}\alpha_{22} , \qquad (4.13)$$

$$\alpha_{21} = \eta' - \eta" + \mu\left[Gh(\eta' + \eta") - \frac{\ell}{\Omega}\eta'\eta" - G^2\Lambda\right] ,$$

$$\alpha_{22} = 1 + \mu\left(Gh - \frac{\ell}{\Omega}\eta"\right) .$$

The parameter G which appears in these equations is equal to

$$G = \frac{F_2(\bar{x})}{F_1(\bar{x})} ,$$

where $\bar{x}$ is the coordinate of the beam and the parameters h, Λ, η' and $\eta"$ are determined by the equations

$$h = \varepsilon F_0(\bar{x})(\sin \phi'' - \sin \phi') ,$$

$$\Lambda = \varepsilon^2 \Omega F_0^2(\bar{x}) \int_{\phi'}^{\phi''} \cos^2 \phi \, d\phi =$$

$$= \frac{1}{2} \varepsilon^2 \Omega F_0^2(\bar{x})(\ell + \sin \phi'' \cos \phi'' - \sin \phi' \cos \phi') ,$$

$$\eta' = \varepsilon \Omega F_0(\bar{x})(G - \alpha') \cos \phi' ,$$

$$\eta'' = \varepsilon \Omega F_0(\bar{x})(G - \alpha'') \cos \phi'' .$$

For equilibrium particles, as follows from Eq. (2.35), the coefficient $h = 1$. The matrix elements are obviously dependent on the shape of the entrance and exit apertures, and this dependence is communicated through the coefficients α' and α''.

Equations (4.13) determine the vertical motion in a microtron with the exception of the first few revolutions. From Eqs. (4.8) and (4.11) it is possible to obtain the matrix (δ_{ik}) which determines the focusing ability of the cavity as a whole as an intermediate step in calculating the matrix (α_{ik}). The elements of this matrix are

$$\delta_{11} = 1 - \mu \left(Gh - \frac{\ell}{\Omega} \eta' \right), \quad \delta_{12} = \mu \frac{\ell}{\Omega} ,$$

$$\delta_{21} = \alpha_{21} , \quad \delta_{22} = \alpha_{22} .$$

(4.14)

In utilizing the matrix (δ_{ik}) in an arbitrary case of focusing in alternating fields, such as in a linac, for instance, let us note that the values ε and Ω enter Eqs. (4.13) and (4.14) only as a product $\varepsilon\Omega$, which is determined by the electric field at the axis of the cavity:

$$\varepsilon\Omega = \frac{eE}{mc\omega} = \frac{1}{2\pi} \frac{eE\lambda}{mc^2} \quad (4.15)$$

One must keep in mind that the matrix (δ_{ik}) can only be used for relativistic particles, the trajectory of which make small angles with the median plane.

We point out that the determinants of all the matrices are equal to unity within the accuracy accounted by us of the terms of the order μ, which is as it should be in the general case (see Appendix II).

SECTION 3 - CALCULATIONS FOR RADIAL FOCUSING

Radial motion is calculated by the same general method used in the previous section. First of all, we calculate the motion inside the cavity where the apertures are replaced by equivalent thin lenses; the matrix obtained is then multiplied by the matrix which describes motion in a uniform magnetic field outside the cavity.

In dimensionless parameters the linearized equations which determine motion of particles along the median plane of the cavity have the following form [see Eq. (3.14) and Fig. 4.5]:

$$\frac{dx}{d\phi} = \frac{q}{\Gamma} ,$$

$$\frac{dq}{d\phi} = [\epsilon\Omega\, F_1(x)\sin\phi + \Omega]v , \qquad (4.16)$$

$$\frac{d\Gamma}{d\phi} = \epsilon\Omega v\, F_0(x)\cos\phi .$$

For solving this system of Eqs. (4.16) we consider that the particle moves close to the axis of the cavity and thus, we can say

$$F_0(x) = 1 , \quad F_1(x) = F_1'(0)x .$$

Further, we consider that the particle velocity is close to the speed of light, and the angle of inclination of the trajectory to the axis of the cavity is small so that we can make $v = 1$.

As a result, we obtain the linear nonhomogeneous system of equations

$$\frac{dx}{d\phi} = \frac{q}{\Gamma} ,$$

$$\frac{dq}{d\phi} = \epsilon\Omega\, F_1'(0)x\sin\phi + \Omega , \qquad (4.17)$$

in which we must consider $\Gamma = \Gamma_0 + \epsilon\Omega \cdot \sin\phi$. The nonhomogeneity of this system (the component Ω on the right-hand side of the second equation) is caused by the magnetic field. The solution of this system is sought in the form of the expansion

$$x = x^{(0)} + \mu x^{(1)} + \ldots ,$$

$$q = \frac{1}{\mu}q^{(-1)} + q^{(0)} + \mu q^{(1)} + \ldots , \qquad (4.18)$$

$$\Gamma = \frac{1}{\mu}\Gamma^{(-1)} + \Gamma^{(0)} + \ldots .$$

Such a form of series is due to the fact that in a uniform magnetic field the electron returns to the point of origin independently of the exit angle from the cavity. Thus, on acceleration the coordinate changes are small at the time when the momentum q and energy Γ receive approximately equal increments at each revolution, i.e., for a constant coordinate they increase in proportion to the number of orbits. Taking into account that the small parameter μ is inversely proportional to the number of orbits [see Eqs. (4.5) and

(2.12)] we obtain the first terms of the expansion (4.18).

The solution to the systems of Eqs. (4.17) is obtained by the method of successive approximations, substituting expansion (4.18) into it and equating the coefficients for the same power of μ. In the result we obtain the matrix elements $(\tilde{\beta}_{ik})$ which relate the vectors $\begin{pmatrix} x \\ q \end{pmatrix}$ for radial motion inside the cavity in the form*

$$\tilde{\beta}_{11} = 1 - \mu\varepsilon\, F_1'(0)(\sin\phi'' - \sin\phi' - \ell\cos\phi') \quad,$$

$$\tilde{\beta}_{12} = \mu\,\frac{\ell}{\Omega} + \mu^2\,\frac{\varepsilon}{\Omega}\{[F_1'(0) - 1](\cos\phi' - \cos\phi'' - \ell\sin\phi') +$$

$$+ F_1'(0)(\cos\phi' - \cos\phi'' - \ell\sin\phi'')\} \quad, \qquad (4.19)$$

$$\tilde{\beta}_{21} = \varepsilon\Omega\, F_1'(0)(\cos\phi' - \cos\phi'') \quad,$$

$$\tilde{\beta}_{22} = 1 + \mu\varepsilon\, F_1'(0)(\sin\phi'' - \sin\phi' - \ell\cos\phi'') \quad.$$

The matrices which characterize the action of the flight apertures are completely analogous to the matrices (4.11) with the exception that the coefficients α' and α'' in them are replaced by coefficients $\tilde{\alpha}'$ and $\tilde{\alpha}''$ which enter into the relationship

$$\frac{\partial E_x}{\partial x} = -\tilde{\alpha}\,\frac{\partial E_y}{\partial y} \quad, \qquad (4.20)$$

which is analogous to relationship (4.9).

In order to find the matrix which corresponds to the motion outside the cavity let us note that the coordinate and the momentum of the particle at the entrance to the cavity will be exactly the same as if it were moving "backwards" through the cavity in the absence of a high frequency field: such is the circular motion in a uniform magnetic field. Considering $\varepsilon = 0$ in the system of Eqs. (4.17) and substituting $-\phi$ for ϕ we obtain the matrix

$$\begin{pmatrix} 1 & -\mu\,\frac{\ell}{\Omega} + \mu^2\,\frac{\ell}{\Omega}\,\varepsilon\,(\sin\phi'' - \sin\phi') \\ 0 & 1 \end{pmatrix} \qquad (4.21)$$

From expressions (4.19) and (4.21) we obtain the matrix $(\tilde{\alpha}_{ik})$ which corresponds to a complete revolution of the electron (for this the start and the finish of the revolution is picked the same as for vertical motion, that is to say, near the entrance to the cavity: see Fig. 4.5). Its elements have the form

---

*The values which make up the equations of radial motion and which are analogous to those used for calculating vertical motion will be designated by the same letters with a wavy line over them.

$$\tilde{\alpha}_{11} = 1 - \mu \left(\tilde{G}h - \frac{\ell}{\Omega}\tilde{\eta}''\right) \quad ,$$

$$\tilde{\alpha}_{12} = \mu^2\left[\frac{\ell^2}{\Omega^2}\tilde{\eta}'' + \frac{2\varepsilon}{\Omega}(\tilde{G} - \tfrac{1}{2})(\cos\phi' - \cos\phi'' - \ell\sin\phi'')\right] \quad ,$$

$$\tilde{\alpha}_{21} = \tilde{\eta}' - \tilde{\eta}'' \quad , \tag{4.22}$$

$$\tilde{\alpha}_{22} = 1 + \mu\left(\tilde{G}h - \frac{\ell}{\Omega}\tilde{\eta}''\right) \quad ,$$

where

$$\tilde{G} = F_1'(0), \quad h = \varepsilon(\sin\phi'' - \sin\phi') \quad ,$$

$$\tilde{\eta}' = \varepsilon\Omega(\tilde{G} - \tilde{\alpha}')\cos\phi' \quad , \quad \tilde{\eta}'' = \varepsilon\Omega(\tilde{G} - \tilde{\alpha}'')\cos\phi'' \quad .$$

The above matrices were obtained by solving the homogeneous part of the system (4.17). In order to obtain the general solution of the whole system it is necessary to add the particular solution which satisfies the zero initial condition. The particular solution of the system (4.17) in the first approximation is equal to

$$x = \mu\frac{\phi^2}{2} \quad , \quad q = \Omega\phi \quad . \tag{4.23}$$

Investigating motion outside the cavity, as previously (backward motion with $\varepsilon = 0$), we obtain

$$x = \mu\frac{\phi^2}{2} \quad , \quad q = -\Omega\phi \quad . \tag{4.24}$$

Taking into account relationships (4.23) and (4.24) we discover that the motion on the complete revolution is described as before by the matrix $(\tilde{\alpha}_{ik})$.

The matrices (4.13) and (4.22) obtained above, completely determine the properties of vertical and radial motion. Before we begin to analyze these properties, let us calculate some of the parameters which make up these matrices.

The focusing action of the cavity depends on the flight phase. Inasmuch as the amplitude of the stable phase oscillations is rather small, we can substitute, in Eqs. (4.13) and (4.22), the equilibrium values for phase and energy. By doing so, we neglect the effect of phase oscillations on focusing. Such an effect can only become troublesome if there is a resonance interaction between the phases and the vertical oscillations.[14]

By substituting equilibrium values for phase and energy and making use of Eqs. (2.12) and (2.35) we obtain

$$h = 1 \quad , \quad \mu = \frac{\Omega}{\Gamma'} = \frac{1}{n} \quad ,$$

$$\phi' = \phi_s - \frac{\ell}{2} \quad , \quad \phi'' = \phi_s + \frac{\ell}{2} \quad , \tag{4.25}$$

where n is the number of orbits and ϕ_s is the equilibrium phase.

Let us now calculate parameters G and $\tilde{G}$. Relationship (4.3) gives the following connection between them

$$G + \tilde{G} = 1 \quad . \tag{4.26}$$

In a circular and in a square cavity, both of these values are equal to

$$G = \tilde{G} = \tfrac{1}{2} \quad . \tag{4.27}$$

For a rectangular cavity with a radial dimension "a" and a vertical dimension "b", from Eqs. (3.16) and (4.3) we find

$$G = \frac{a^2}{a^2 + b^2} \quad , \quad \tilde{G} = \frac{b^2}{a^2 + b^2} \quad . \tag{4.28}$$

We will now calculate coefficients α and $\tilde{\alpha}$ which characterize the shape of the apertures and which are introduced into Eqs. (4.9) and (4.20). These relationships can be derived from several limiting cases: for circular apertures, for infinitely long vertical or horizontal slits, for elliptical apertures in an infinitely thin wall, and others. We can thus accept these relationships (at least in the first approximation) in all cases where the flight aperture is symmetric with the vertical or horizontal planes, keeping in mind that usually these relationships must be used for slit type and circular apertures for which they are wittingly accurate. Equation div $E = 0$ and Eqs. (4.9) and (4.20) lead to the relationship

$$\alpha + \tilde{\alpha} = 1 \quad . \tag{4.29}$$

Coefficient α can be determined from the following considerations. Close to the y axis, equipotential lines have the shape of an ellipse for which the ratio of the horizontal to the vertical axis is equal to c/d, where c is the horizontal and d the vertical dimension of the flight aperture. From this we find that the value of the parameter α is equal to

$$\alpha = \frac{c^2}{c^2 + d^2} \quad . \tag{4.30}$$

Actually, for a vertical slit $\alpha = 0$, for circular or square apertures $\alpha = 0.5$, and for horizontal slits $\alpha = 1$. These results are accurate.

From Eq. (4.30) it can be seen that focusing depends only on the shape of the aperture and not on its absolute size. This result seems to be quite natural since we replaced the apertures by equivalent thin lenses.

Let us note that symmetry of the flight apertures and the body of the cavity relative to the vertical and horizontal planes leads to the fact that vertical and radial motion in the linear approximation are apparently independent of

MICROTRON 83

each other. We have limited ourselves to the investigation of this case, and if this symmetry is destroyed this independence no longer exists.

SECTION 4 - ANALYSIS OF VERTICAL AND RADIAL MOTION

Stability of cyclic motion is characterized by the trace of the matrix: motion is stable if the absolute value of the trace is less than 2 [see Eq. (2.16)].

Let us calculate the trace of the matrix for vertical motion. At the outerorbits (n >> 1), elements which are of the order 1/n can be ignored. In this case, as can be seen from Eq. (4.13), the trace of the matrix does not depend on the number of orbits and is equal to

$$S = 2 - \frac{2\pi}{\Omega} (\eta'' - \eta') \quad . \tag{4.31}$$

Utilizing relationship (4.25) and the condition $\tan \phi_s = 1/\pi$, which is true for the optimum equilibrium phase [see Eq. (2.24)], we can represent the value S in the form

$$S = 2(1 + G) - (\alpha' + \alpha'') + \pi(\alpha'' - \alpha') \cot \ell/2 \quad . \tag{4.32}$$

This expression determines stability of vertical motion for various shapes of cavities and flight apertures. In Table 4.1 we present values of S for several given parameters of the cavity and apertures.

TABLE 4.1 The Value of S for Various Parameters of a Cylindrical Cavity

Cavity Thickness, ℓ	⊙▶⊙	⊂▶⊂	⊙▶❘	⊂▶⊙	⊂▶❘
1.0	2	1	-0.37	-1.37	-3.75
1.2	2	1	0.20	-0.80	-2.60
1.4	2	1	0.64	-0.36	-1.72

In the case where the cavity and the flight apertures possess symmetry of revolution relative to the axis we have

$$G = \alpha' = \alpha'' = \tfrac{1}{2} \quad \text{and} \quad S = 2 \quad .$$

This means that the conditions of motion exist on the boundary of the stable region. As is shown in Appendix II, when the condition S = 2 is completely satisfied, the solution has the form [see Eq. (II.10)] $z_n \sim n$, which corresponds to a trajectory which has the shape of a spiral line increasing in radius, as mentioned in Section 1 of this Chapter. Such motion would indicate strong instability.

However, by taking into account terms of the order $1/n$ in Eq. (4.13) we have

$$S = 2 - \frac{1}{n}\frac{\pi \Lambda}{2\Omega} \quad , \tag{4.33}$$

i.e., $S < 2$ and the motion is stable. As can be seen from Eqs. (4.6) and (4.13), terms of the order $1/n$ appear as a result of a change in the coordinates inside the cavity.

Since the entrance aperture focuses the particles, this change in the coordinate makes the defocusing weaker and brings about stable motion.

However, this focusing is very weak; actually, the system stays on the border of stability, since $S \to 2$ as n increases. As we have noted previously, vertical motion in the case of circular symmetry was calculated by Bell[61]; expression (4.33) agrees with the formula derived by him. Bell showed that in this case oscillatory motion takes place but the period grows in proportion to $\sqrt{n}$, and the amplitude of oscillations increases in proportion to $\sqrt[4]{n}$.

Increase in the amplitude of oscillations causes particles to be lost. Furthermore, various weak disturbances, which are usually insignificant when there is a high degree of stability, in this case will bring about additional particle losses.

As an example, let us take the case in which the cavity is not accurately positioned, for instance, when the axis of the cavity is inclined through a small angle ϑ to the horizontal plane. Let the coordinate system for the cavity be the same as previously, i.e., the y axis is inclined through an angle ϑ to the median plane of the magnet. In this case matrices (4.8) and (4.11) which correspond to motion through the cavity, in the first approximation remain undisturbed. Motion outside the cavity can be calculated if the following substitution was made into matrix (4.12) which determines motion in a homogeneous magnetic field

$$z \to z + \ell\vartheta \quad , \qquad p \to p + (\Gamma' + \Omega)\vartheta \quad ,$$

$$z^* \to z^* \quad , \qquad p^* \to p^* + (\Gamma' + \Omega)\vartheta \quad .$$

The result is

$$\begin{pmatrix} z_{n+1} \\ p_{n+1} \end{pmatrix} = \begin{pmatrix} \alpha_{11} & \alpha_{12} \\ \alpha_{21} & \alpha_{22} \end{pmatrix} \begin{pmatrix} z_n \\ p_n \end{pmatrix} + \begin{pmatrix} \frac{2\pi\vartheta}{\mu} \\ 0 \end{pmatrix} \quad , \tag{4.34}$$

where the matrix (α_{ik}) represents undisturbed motion.
The difference equation has the following form

MICROTRON

$$z_{n+1} = S_n z_n - z_{n-1} + 2\pi\vartheta \left(1 - G + \frac{\ell}{\Omega}\eta''\right) \quad . \qquad (4.35)$$

For circular symmetry the trace S_n is determined by expression (4.33). The particular solution for Eq. (4.35) according to Eq. (II.16) in Appendix II has the form

$$z_n = \frac{2\Omega}{\Lambda}\vartheta n \quad . \qquad (4.36)$$

In this way, vertical displacement of the beam takes place in proportion to the number of orbits. If $\vartheta = 10'$ and n = 30, then the displacement $\Delta z \simeq 0.2$, which could cause an almost total loss of the beam.

Other disturbances, such as, for instance, a nonuniform magnetic guide field, the effect of the magnetic field induced by the cathode heating current, etc., lead to an expression similar to Eq. (4.36), i.e., they will have a strong effect in the case of a circular cavity with round flight apertures.

We will now investigate the properties of vertical motion in an arbitrarily shaped cavity and flight apertures. In general, the value S determined by relationship (4.32) is not close to 2 (see Table 4.1), and depending on the choice of parameters may lie either inside or outside the stable region. It is interesting to note that S is determined only by the geometry of the cavity and the flight apertures, and does not depend on the electric and magnetic fields.

This property can be explained in the following way. The cavity is equivalent to a lens, the focusing of which is proportional to the high frequency field. The length of the orbits is equivalent to the distance between these lenses and inversely proportional to the constant magnetic field value.

It is easy to see by investigating the sequence of the focusing lenses that the expression for S will contain the product of the power of the lenses and the distance between them which is proportional to ε, and, for a constant geometry cavity, is fixed by the main parameters of the microtron acceleration mode [see Eqs. (2.7) and (2.35)].

The dependence of focusing only on the cavity geometry is important from a practical standpoint, since for a given cavity geometry, it is possible to vary the parameters of the acceleration mode within wide limits, i.e., vary the energy gain per revolution and the strength of the magnetic field. Vertical stability with this is not destroyed.

Let us now determine the geometric parameters of the cavity for which vertical focusing is optimal. As we have already pointed out in Chapter II, the value of S for this is equal to zero. On Fig. 4.6 are shown lines of S = constant as functions of the shapes of the flight apertures (coefficients α' and α'' are on the abscissa), for various widths of the cavity ℓ, and for various values of the parameter G which depends on the shape of the cavity. The values of ℓ and G are usually dictated by the conditions of injection. As can be seen from Fig. 4.6, regardless of any values of ℓ and G that may be encountered in practice, it is always possible to

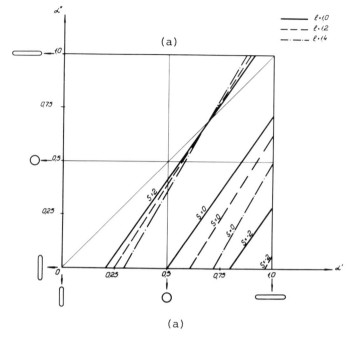

(a)

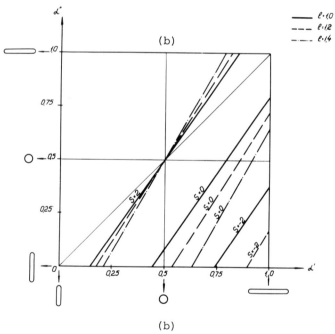

(b)

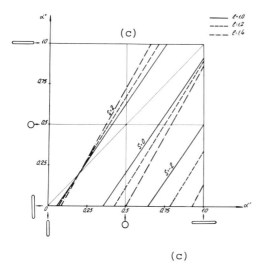

(c)

Fig. 4.6 Properties of vertical motion with variously shaped flight apertures and different thicknesses of the accelerating cavity: (a) Rectangular cavity, elongated in the horizontal direction; (b) Cylindrical cavity; (c) Rectangular cavity, elongated in the vertical direction.

insure optimum vertical focusing by choosing apertures of the proper shape.

Let us discuss now an example. Assume that we use a cylindrical cavity, i.e., G = 0.5. If instead of circular apertures, we were to use horizontal elongated slits at the entrance to and the exit from the cavity, then $\alpha' = \alpha'' = 1$. In this case, as follows from Eq. (4.32), the value of S is equal to unity and does not depend on the width of the cavity. The period of vertical oscillations is equal to six revolutions and, on the outer orbits, does not change with increase in particle energy.

From expression (4.32) it can be seen that focusing is strongly affected by changes in the shape of the flight apertures. The more the entrance aperture is spread out in the horizontal direction ($\alpha' \to 1$) and the more the exit aperture is spread out in the vertical direction ($\alpha'' \to 0$), the stronger is the focusing action of the cavity. From Fig. 4.6 and also from Table 4.1 it can be seen that by changing the shape of the flight apertures, the conditions of motion can pass from a region of defocusing (S > 2) into a region of stability ($|S| < 2$), whereas for crossed slits it goes into a region of instability which corresponds to the value of S < -2.

Optimum focusing (S = 0) can be achieved by combining apertures of various shapes. For example, a horizontal slit at the entrance and a circular aperture at the exit from the cavity, or a circular aperture at the entrance and a vertical slit at the exit appear to be optimum for focusing. It is also possible to have other combinations. For this reason the conditions of vertical stability leave a considerable latitude in the choice of the shape of apertures. In each real case, the shapes of the apertures should be selected by taking into account the conditions of radial motion as well as the conditions of injection.

The shape of the cavity also affects focusing. As follows from Eqs. (4.28) and (4.32), the parameter G, and with it the value S, are larger in a cavity which is spread out in the horizontal direction, i.e., vertical focusing becomes weaker. And, conversely, increasing the vertical dimension of the cavity brings about stronger focusing. This property can be explained by the fact that the horizontal component of the alternating magnetic field which defocuses the particles increases as the horizontal dimension of the cavity is made larger.

It is evident from Eqs. (4.28) and (4.30) that parameters G, α' and α'' are determined by the same type of expressions and depend only on the relationship of the vertical and horizontal dimensions of the cavity (G) and the apertures (α' and α''). If this ratio is the same for the cavity and both apertures, then $G = \alpha' = \alpha''$ and $S = 2$.

Focusing also depends on the thickness of the cavity ℓ. For a larger ℓ the alternating field must be smaller in order that the energy gain per revolution remain unchanged. This weakens the focusing action of the cavity. As can be seen from Fig. 4.6, the upper boundary of the stable region ($S = 2$) stays almost in the same position with increase in ℓ while the lower boundary ($S = -2$) is shifted downward. The result is that the region of stability increases somewhat. Equation (4.32) shows that for the identical shape aperture ($\alpha' = \alpha''$), the value S does not depend on the thickness of the cavity.

Let us now determine certain characteristics of vertical motion. As we have already indicated, after several initial revolutions it is possible to neglect terms of the order $1/n$ which enter into expression (4.31). Then the trace of the matrix is expressed by Eq. (4.32) and does not depend on the number of orbits. Consequently, the particles oscillate vertically at a constant frequency and amplitude. From Eqs. (4.13) and (II.7), p_n and z_n are obtained in the form

$$p_n = C \cos(\nu n + \chi) ,$$

$$z_n = \rho C \cos(\nu n + \chi + \zeta) , \qquad (4.37)$$

where

$$\cos \nu = \frac{S}{2} , \quad \rho = -\frac{2}{\Omega(2 - S)} ,$$

$$\cos \zeta = -\frac{\sqrt{2 - S}}{2} , \quad \sin \zeta = \frac{\sqrt{2 + S}}{2} .$$

C and χ are arbitrary constants which are determined by the initial conditions.

MICROTRON

From expression (4.37) it can be seen that all the characteristics of vertical motion at the outer orbits are determined by the value S and consequently the relationship between the amplitudes of oscillation of z_n and p_n depend also on the magnetic field parameter Ω. In the z_n, p_n plane the points form an ellipse:

$$z_n^2 - \frac{2\pi}{\Omega} z_n p_n + \frac{4\pi^2}{\Omega^2 (2 - S)} p_n^2 = \frac{2 + S}{4} z_{max}^2$$

$$p_{max} = \frac{\Omega}{2\pi} \sqrt{2 - S} \, z_{max}$$

(4.38)

where z_{max} and p_{max} are the amplitudes of the oscillations. For a given initial value of z_n and p_n the amplitude of oscillations becomes larger as the conditions of focusing approach the boundary of stability, i.e., as $|S|$ approaches 2. The width of the beam in the direction of the z axis is limited by the dimensions of the flight apertures and thus, the closer the system is to the boundary of stability the larger portion of the beam is lost on the walls of the cavity.

Let us evaluate the values characterizing vertical focusing for the optimum value of S = 0. Assuming $z_{max} = 0.25$ (in practice, the entrance aperture is approximately this size), from the second equation (4.38) we will find the maximum possible angular spread of the beam in the vertical plane

$$\Delta\theta_{max} = \frac{2}{n\Omega} p_{max} = \frac{\sqrt{2}}{\pi n} z_{max} \approx \frac{0.1}{n} \; . \qquad (4.39)$$

On the tenth orbit $\Delta\theta_{max} = 0.01 = 0°,6$, on the 30th orbit $\Delta\theta_{max} = 0.003 = 0°,2$.

Let us evaluate the effect of disturbances. We have established above that even a slight inclination of the cavity will strongly disturb motion if $S \approx 2$. Let us now assume that the apertures in the cavity have some other shape for which motion is highly stable. Vertical displacement of the beam under the action of the cavity inclination is obtained as a particular solution of the difference equation (4.35) in the form of a constant

$$\bar{z} = \frac{2\pi\vartheta}{2 - S} \left(1 - G + \frac{\ell}{\Omega} \eta''\right) \; . \qquad (4.40)$$

Let us investigate an actual case: assume a circular cavity with a thickness $\ell = 1.2$, and two horizontal slits. In this case $S = 1$, $\varepsilon = 0.95$, and $\bar{z} = \vartheta$. If we assume that the allowable displacement $\bar{z} = 0.05$, then the allowable cavity inclination is approximately 3°. This displacement of the beam does not depend on the number of orbits.

We can see that replacing round apertures by horizontal slits can substantially decrease the disturbance introduced by angular misalignment of the cavity. In a similar way, disturbances due to other factors are also weakened. In Section 4 of Chapter VII, it is shown that this theoretical result is substantiated experimentally.

Let us now evaluate the properties of radial motion. In the above equations (see Section 3 of this Chapter), the values x and q correspond to that point on the orbit which is placed at the entrance to the cavity. Radial motion determined by this method is not directly connected to phase motion which mainly affects the longitudinal momentum and the radius of the orbit.

Let us compare the matrix for radial motion (4.22) with the matrix for vertical motion (4.13). The expression for elements $\tilde{\alpha}_{21}$ and $\tilde{\alpha}_{22}$ in the matrix (4.22) can be obtained if we replace parameters G and α by similar parameters $\tilde{G}$ and $\tilde{\alpha}$ for radial motion in the corresponding elements of matrix (4.13). These results become clear if we consider that these matrix elements are determined only by the focusing properties of the cavity.

Qualitative differences in the radial and vertical motions appear in elements $\tilde{\alpha}_{12}$ and α_{12}. For vertical motion $\alpha_{12} \approx 2\pi/\Omega$ = const, which corresponds to motion outside the cavity along a spiral line. The element $\tilde{\alpha}_{12}$ is proportional to μ since the particle moving around in the uniform field returns to the point of origin and element $\tilde{\alpha}_{12}$ is not equal to 0 because of the finite thickness of the cavity.

As we have already mentioned, the determinants of matrices (4.22) and (4.13) are equal to unity. By taking this circumstance into account, the values of elements α_{12} and $\tilde{\alpha}_{12}$ immediately determine the possible trace of these matrices Thus, for vertical motion, trace S can take on any value, whereas the trace $\tilde{S}$ corresponding to radial motion, differs from 2 only by small terms of the order μ^2. All these considerations clarify the sense of Eq. (4.22).

The circumstance that $2 - \tilde{S}$ is of the order μ^2, essentially means that there is an absence of radial focusing in a microtron. The exact value $\tilde{S} = 2$ can be obtained if we neglect terms of order μ and μ^2 in Eqs. (4.22). In that case (see Appendix II),

$$x_n = x_1 = \text{const} , \quad q_n = q_1 n . \qquad (4.41)$$

The angle of inclination of the trajectory to the axis of the cavity is proportional to $q/\Gamma \sim q/n = q$, i.e., it would remain constant. The exact equality $\tilde{S} = 2$ and the resulting conditions correspond to the limiting case discussed in Section 1 of this Chapter (zero thickness cavity).

Since the value $\tilde{S} = 2$ is particular, in order to obtain an exact solution it is necessary to take small deviations of $\tilde{S}$ from 2 into account. This has been accomplished in Ref. 14, in which it was shown that real motion in a uniform magnetic field is stable. The coordinate x and the angle of inclination of the trajectory to the axis of the cavity slowly diminish and, in contrast to vertical motion, radial motion does not have an oscillating characteristic. This variation of the coordinate x determined by the action of the cavity is relatively small and a considerably larger variation of x can be caused by the action of a nonuniform magnetic field (see Section 5 of this Chapter). However, field nonuniformity usually has a smooth characteristic and thus displaces the

beam as a whole, so the decrease in beam radial dimensions and angular spread under the action of the alternating field of the cavity takes place as before.
Equation (4.22) allows us to make certain conclusions about angular spread of the beam. Comparing Eqs. (4.22) and (4.13) and taking into account the relationships (4.26) and (4.29) we have

$$\tilde{\eta}' = -\eta' \quad , \quad \tilde{\eta}'' = -\eta'' \tag{4.42}$$

and consequently,

$$\tilde{\alpha}_{21} = -\alpha_{21} = \frac{\Omega}{2\pi}(2-S) \quad . \tag{4.43}$$

This relationship has an obvious physical meaning. It means that focusing in any direction is necessarily accompanied by defocusing in the transverse direction. Such action of the cavity is exactly analogous to the action of a single quadrupole lens, electrical or magnetic. Only in the case where

$$\tilde{\alpha}_{21} = -\alpha_{21} \approx 0$$

does there exist weak focusing in both directions due to second-order terms associated with coordinate changes during motion inside the cavity. In this case, the cavity acts as two quadrupole lenses which produce alternating focusing in both directions.
From relationship (4.43) it follows that for $S < 2$, when vertical motion is stable we have $\tilde{\alpha}_{21} > 0$, so that there is radial defocusing. The smaller the value S, i.e., the stronger the cavity focuses in the vertical direction, the more there is radial defocusing. For this reason there is no aperture shape which is optimal from all standpoints: in each real case the aperture should be selected by taking into account the requirements placed on the beam. Let us note that a combination of two horizontal slits insures both vertical and radial focusing and appears to be the most practical.
It must be mentioned here, that in certain cases, for instance in the second type of acceleration, the orbits are displaced horizontally to the edge of the aperture. When this occurs, the vertical component of the electric field decreases relative to that which exists at the axis of the aperture and is determined by Eq. (4.9). The focusing force also diminishes accordingly. In such a nonparaxial case, the above equations cannot be used and vertical motion must be studied empirically.

SECTION 5 - PARTICLE MOTION IN A SLIGHTLY NONUNIFORM MAGNETIC FIELD

Previously we have assumed that the magnetic field in a microtron was uniform and constant in time.

Because of the large inductance of the magnet, the rise time of the magnetic field is considerably larger than the time required to accelerate particles to the final energy. Consequently, motion in a changing field differs little from motion in a uniform field since at each moment the energy of the accelerated particle and the value of the equilibrium phase correspond to the instantaneous value of the field. Inasmuch as the equilibrium phase can vary within predetermined limits (see Section 4 of Chapter II), then for the same conditions the allowable change in the magnetic field amounts to several percent. However, the induced eddy currents from the changing magnetic field destroy the motion of the particles.* In order to avoid this and also maintain the constant energy of the accelerated particles, it is common to stabilize the current in the field to an accuracy of several tenths of a percent.

Let us now investigate the effect of field nonuniformity on the motion of particles. Nonuniformity affects all forms of motion: phase, vertical, and radial. If the nonuniformity is symmetric relative to the plane $z = 0$, i.e., if there exists a median plane of the magnet which coincides with the plane of the orbit of the electrons, then the effect of the nonuniformity on the vertical motion is limited to a change in the frequency of the vertical oscillations. In order to take this into account rigorously, it is necessary to change matrix (4.12) which describes motion outside the cavity. We will evaluate this disturbance in an approximate way, by comparing the action of the cavity on the vertical motion with the action of the nonuniform magnetic field.

The action of the cavity can be characterized by introducing $n_{equiv.}$, the index of the equivalent magnetic field which insures the same frequency of vertical oscillations as a microtron cavity:

$$n_{equiv.} = \left(\frac{\nu}{2\pi}\right)^2 , \qquad (4.44)$$

where ν is expressed by Eq. (4.37). For the case of optimal focusing $\nu = \pi/2$ and $n_{equiv.} = 0.062$. The effect of the nonuniformity on each orbit is determined by the field index

$$n = -\frac{r}{H}\frac{\partial H}{\partial r} , \qquad (4.45)$$

where r is the radius-vector as measured from the center of the orbit. In microtron magnets the maximum value $|n|$ is

---

*The effect of eddy currents on particle focusing can be easily observed by rapidly varying the field close to its equilibrium value. When the field is lowered there is an intensification of the vertical focusing since the profile of the field becomes blown-up. On increasing the magnet current the beam disappears because of the defocusing action of the field rising toward the periphery. It is interesting to note that supplementary shimming of the field during changing of the current played a large role in the discovery of He^3 on the cyclotron by L. Alvarez.[91]

usually between 0.001-0.005. By equating $n_{equiv.}$ with n, we see that in the given case the effect of field nonhomogeneity is insignificant. This effect can become significant only in two cases: when the frequency of vertical oscillations is small, and also when the nonuniformity has a special shape and amounts to several percent (the latter case is investigated at the end of this Section).

Vertical motion is more strongly disturbed if the median plane is distorted or displaced relative to the plane of the electron orbit. Such a nonuniformity leads to the vertical displacement of the orbits but in practice it can be neglected, if focusing is optimum.

Let us now go on to radial and phase motion. In the ideally uniform field, the particle moves around and after a complete revolution returns exactly to the point of origin with the same horizontal momentum; the time of revolution is exactly proportional to the total particle energy, thus the equilibrium phase is constant. In a slightly nonuniform field, the particle trajectory is slightly distorted. After completing the revolution, the particle returns to the point slightly displaced from the origin, and the angle of inclination of the trajectory at this point is also slightly changed. The time of revolution is also changed which causes a shift in the equilibrium phase.

The indicated effects were initially calculated by V. P. Bikov[9] who evaluated radial drift of the orbit and displacement of the equilibrium phase in a slightly nonuniform field. In calculating radial drift the action of the cavity was not taken into account. Apparently, however, the alternating field of the cavity can somewhat change the drift along the common diameter of the orbit (we will call it the longitudinal drift), and appreciably affects the drift transversely to the common diameter. However, transverse drift has a weak effect on the operation of a microtron (see Section 1 of this Chapter) and will not be calculated here, while longitudinal drift caused by a nonuniformity cannot be appreciably changed by the field in the cavity (as compared, for instance, with vertical and phase drift). In order to simplify the following derivations, we will neglect the effect of the cavity on the longitudinal drift of the orbit (the solution of this problem without neglecting the effect of the cavity is given in Ref. 14).

Let us investigate particle motion on the median plane z = 0. In this plane the field has only a vertical component H_z, which is equal to

$$H_z = H[1 + h(x,y)] \;, \qquad (4.46)$$

where H is the field in the center of the magnet and the function $h(x,y)$ satisfies the inequality $|h(x,y)| \ll 1$ (we assume a slight nonuniformity). Making use of the dimensionless variables introduced in Section 1, Chapter II, the equation of plane motion can be written in the form

$$\frac{du}{d\phi} = \frac{\Omega}{\Gamma} v[1 + h(x,y)] ,$$

$$\frac{dv}{d\phi} = -\frac{\Omega}{\Gamma} u[1 + h(x,y)] .$$
(4.47)

This system of equations can be approximately solved by the following method (see Ref. 14). Let us introduce a polar angle ϑ and also θ as the angle of inclination of the trajectory to the axis y (see Fig. 4.7). Thus, we have

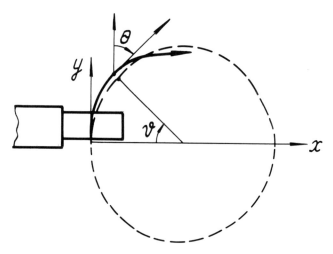

Fig. 4.7 For calculating motion in a nonhomogeneous magnetic field.

$$u = \frac{dx}{d\phi} = \beta \sin \theta , \quad v = \frac{dy}{d\phi} = \beta \cos \theta .$$
(4.48)

By combining relationships (4.47) and (4.48) we obtain the equation

$$\frac{d\theta}{d\phi} = \frac{\Omega}{\Gamma} [1 + h(x,y)] .$$
(4.49)

For undisturbed motion along a uniform field we have

$$\vartheta = \frac{\Omega}{\Gamma} (\phi - \phi_0) .$$
(4.50)

Substituting in Eq. (4.49) the value ϑ in place of ϕ, integrating and making use of h, we obtain the approximate relationships

$$\sin \theta = \sin \vartheta + \cos \int_0^\vartheta h(\vartheta)\, d\vartheta$$

$$\cos \theta = \cos \vartheta - \sin \int_0^\vartheta h(\vartheta)\, d\vartheta \quad ,$$

(4.51)

where $h(\vartheta)$ is the function $h(x,y)$ on the undisturbed circular orbit. Taking into account (4.50) and (4.51) we integrate Eq. (4.48) over ϑ within the limits 0 to 2π. After changing the order to integration in the double integrals, we obtain the expressions

$$x = x_0 - \frac{\beta\Gamma}{\Omega} \int_0^{2\pi} h(\vartheta)\, \sin\vartheta\, d\vartheta$$

$$y = y_0 + \frac{\beta\Gamma}{\Omega} \int_0^{2\pi} h(\vartheta)(1 - \cos\vartheta)\, d\vartheta$$

(4.52)

which determine the point at which the particle will arrive in the time interval equal to a period of revolution in a uniform field. The change in the phase of the entrance of the particle into the cavity in comparison with the undisturbed motion according to the second equation (4.52) is equal to

$$\Delta\phi = -\frac{y - y_0}{\beta} = -\frac{\Gamma}{\Omega} \int_0^{2\pi} h(\vartheta)(1 - \cos\vartheta)\, d\vartheta \quad . \quad (4.53)$$

Relationships (4.48), (4.50) and (4.51) allow us to calculate the increase in the radial component of the momentum per revolution (the comparison is made for $y = 0$, i.e., when the particle enters into the cavity)

$$\Delta q = \Gamma\Delta u = \beta\Gamma \int_0^{2\pi} h(\vartheta)\, \cos\vartheta\, d\vartheta \quad . \quad (4.54)$$

Equations (4.52) to (4.54) were obtained in Ref. 9 by means of more complicated methods. These expressions approximately describe the effect of slight nonuniformity of arbitrary shape and are, essentially, equations of adiabatic approximation. For further calculations we will make use of a suggestion made by S. P. Kapitza, which assumes the nonuniformity can be represented by the second degree polynomial

$$h(x,y) = h_0 + a(x - \rho_N) + by + c(x - \rho_N)^2 + d(x - \rho_N)y + ey^2,$$

(4.55)

where $\rho_N = N + 1$ is the radius of the last, N^{th} orbit [see relationship (2.9) and (2.12)] and the coordinate system is shown on Fig. 4.7. Coefficients of this polynomial are determined by the results of field measurements. Expression (4.55) reflects the lack of symmetry of the microtron orbits and thus, is more convenient than the Fourier expansions normally used in cyclic accelerators with central symmetry.

Further calculations are performed in the following manner. Utilizing the coordinates of the Nth orbit in polynomial (4.55)

$$x = (n + 1)(1 - \cos \vartheta), \quad y = (n + 1) \sin \vartheta$$

and substituting the obtained function $h(\vartheta)$ in expressions (4.52) and (4.53), we obtain the radial drift and the phase displacement on the Nth orbit. By skipping intermediate derivations (they can be found in Ref. 9), we will present only the final expression for the total longitudinal drift obtained by summing on all the orbits while neglecting the action of the cavity:

$$\Delta x_N = \frac{\pi N^2}{3} \left(bN + \frac{d}{4} N^2 \right) . \qquad (4.56)$$

As might be expected from physical considerations, the drift along the common diameter of the orbit depends only on coefficients b and d which determine the asymmetry of the field in that direction. The maximum nonuniformity h_{max}, on the last orbit becomes bN for a linear nonuniformity and $bN^2/2$ for a nonuniformity described by the term $d(x - \rho_N)y$ in Eq. (4.55). If we fix h_{max} and increase the number of orbits, then the total drift determined by expression (4.56) grows in proportion to the square of the number of orbits. Recall that the dimensionless value x is entered in expression (4.56), i.e., the coordinate X expressed in units of $\lambda/2\pi$. Consequently, there is an increase in the number of orbits due to the shorter wavelength (for a fixed diameter magnet) and the radial drift increases in proportion to the number of orbits. For a fixed wavelength and a larger diameter magnet, the drift increases in proportion to the square of the number of orbits. Since the field distribution in the cavity is determined by the dimensionless coordinate x, the effect of the nonuniformity, in both cases, is proportional to N^2.

The change in the radial coordinate determined by the drift of the orbit causes a displacement in the equilibrium phase as a result of the change in the accelerating field along the radius [see Eq. (2.35)]. Taking into account the limiting value of this shift (see Section 4, Chapter II), we find that the maximum possible asymmetry in the field h on the last orbit can be evaluated by the expression

$$h_{max} \approx \frac{1}{3N^2} \qquad (4.57)$$

For this reason the serious technical difficulties take place in the construction of microtrons with large number of orbits.

Indeed, for a 12-orbit microtron, the allowable asymmetry in the field is 0.2-0.3%; such a field uniformity is fairly easy to obtain. In a 30-orbit microtron, h ≈ 0.03% which already means that great accuracy will be needed in the manufacture of the magnet. Further increases in the number of orbits will necessitate the inclusion of supplementary elements which correct the position of the orbit and which eliminate longitudinal radial drift.

The longitudinal orbit drift appears to be the strongest disturbance caused by the nonuniformity. As was shown in Ref. 9, changing the flight phase determined by Eq. (4.53) is considerably less significant. We will not make the derivations, but will explain the physical meaning of this result. Field nonuniformity destroys the proportionality between the total particle energy and time of revolution. This causes a shift in the equilibrium phase which does not depend on the number of orbits, but is determined only by the magnitude h_{max}. In this way, phase motion drift is not added or summed, but there only occurs a shift in the center of the phase oscillations. This is explained by the mechanism of auto-phasing which insures stability of phase motion against all kinds of disturbances.

As we have pointed out previously, magnetic field asymmetry relative to the common diameter of the orbits rather strongly disturbs particle motion and can cause their total loss. If, however, the nonuniformity is such that the field is symmetric relative to this direction (the x-axis on Fig. 4.7), then according to Eq. (4.56) longitudinal drift does not exist. If, in addition, the nonuniformity does not change along the x coordinate (Fig. 4.7), then the transverse drift in such a field is also equal to zero, as can be seen from Eq. (4.52). For simplicity we will call such a two-dimensional field which is symmetric, relative to the y = 0 plane (Fig. 4.7), symmetric, and the corresponding nonuniformity we will also consider as being symmetric.

Investigation of particle dynamics in a microtron with a symmetric magnetic field was performed in detail in Ref. 23; here we will briefly present the main results of this report. Symmetric nonuniformity causes a shift in the equilibrium phase because of the change in the particle time of revolution. This undesirable effect determines the limiting nonuniformity; however, this effect is relatively weak, and the allowable field nonuniformity at the edge of the magnet is approximately 4-5%. Such a nonuniformity noticeably affects vertical and radial motion, and can be utilized for improving the characteristics of the microtron.

Let us investigate the effect of a nonuniformity that decreases field at the edge of the magnet. Such a drooping magnetic field focuses particles in the vertical direction and can be useful in certain modes of acceleration. Besides, the joint action of the magnetic field and the alternating field in a cavity can produce radial focusing of the particles. In order to clarify this result, we have shown on Fig. 4.8 radial motion of particles in a symmetrically drooping field for different exit angles from the cavity. It can be seen from the figure that a positive angle of inclination to the

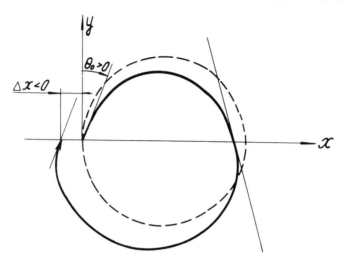

Fig. 4.8 Particle trajectory in a symmetric magnetic field.

axis of the cavity causes a negative displacement in the radial coordinate after completing one revolution.* Since the alternating field in the cavity defocuses the particle in the radial direction that radial motion takes on an oscillating nature, i.e., there exists radial focusing of the particle.

Investigation of the various effects has shown that the frequency of radial oscillations is not great, and consequently are not strong enough to overcome the disturbing action of the various factors. However, the transition to an oscillating regime of motion in itself opens up new possibilities. For instance, it is possible to decrease the radial angular spread of the accelerated beam, by selecting a frequency of oscillations such that on the last orbits there will be an antinode of the radial oscillations. There are also other uses of a drooping symmetric field of this type (see Ref. 23).

---

*In calculating this displacement the first equation in (4.52) was made more accurate by taking into account the dependence of the drift on the initial exit angle. The displacement arising from this is proportional to the value h.

CHAPTER V

CONSTRUCTION OF A MICROTRON

SECTION 1 - GENERAL REMARKS

The microtron builder must solve the same problems which confront a builder of any other modern accelerator. He has to obtain a uniform magnetic field, a high frequency system, and introduce means of injecting and extracting the particles from the accelerator. Finally, the accelerator must have a vacuum chamber, a control system, and power and cooling systems.
We will consider each of these problems, emphasizing those which are specifically concerned with microtrons. In this Chapter we will draw primarily from experience at the Physics Laboratory gained in the construction and utilization of microtrons. This experience is characteristic of a research laboratory: the modern experience of scientific instruments and electrophysical industry has not yet been fully exploited in constructing microtrons. At present most microtrons use an internal Vac-ion pump, placed in the vacuum chamber. This has significantly increased pumping speed and improved the vacuum. A series of design decisions will undoubtedly be reviewed before the construction of the microtron will fully pass into the hands of industry.

SECTION 2 - THE MICROTRON MAGNET

The field in a microtron is obtained by an electromagnet the size and the field of which determine the energy of the beam.*
The magnetic field in a microtron does not exceed 2 to 3 kOe which is substantially lower than that normally used in fixed field cyclic accelerators, cyclotrons for instance, where 18 to 20 kOe is common. The reason for not using higher magnetic fields, say greater than 3 kOe, as explained in Section 2 of Chapter I, is that the amplitude of the high frequency electric field (which in the absolute system of units must be as high as the constant magnetic field) which

---

*In principle it is possible to use a permanent magnet (see Ref. 70). However, in practice this is not reasonable: its weight would be greater than an electromagnet and means for changing its magnetic field (which is imperative for changing the energy of the particles) are very cumbersome.

corresponds to such magnetic fields, will be greater than 10^6 V/cm and can lead to electrical breakdown.

The comparatively small value of the field in the microtron makes the construction of its magnet rather peculiar. The small induction in the poles reduces the cross section of the magnet as a result of the high induction in the yoke. In practice, there are two types of magnets that are practical to build. In the simpler one, the square magnet, the yoke consists of four pillars (see Fig. 1.9); the main advantage of this design is its simplicity and that the walls of the vacuum chamber are accessible for experiments. The square magnet was used on the first Canadian microtron and in a series of other machines.[59,66,76]

A more compact and more advanced construction of the magnet is the armor type (Fig. 5.1). Such a magnet was used

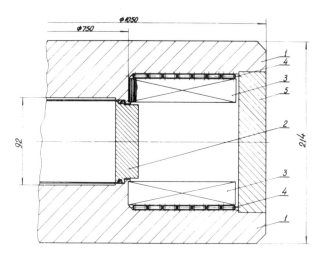

Fig. 5.1 Magnet for the 17-orbit microtron: (1) Pole face; (2) Chamber; (3) Coil; (4) Cooling tube; (5) Iron.

in our 17- and 30-orbit microtrons, the Dubna microtron, and the 2-m London microtron. The main parameters of the magnets are given in Table 5.1.

The field H in the electromagnet is determined by the gap h and the magnetomotive force In. If In is expressed in ampere-turns, H in oersted, and h in centimeters, then

$$H = 0.4 \pi \frac{In}{h} . \qquad (5.1)$$

In the microtron the gap h is approximately equal to the wavelength and in most cases is about 10 cm. Because of the small flux leakage in the magnet and the absence of saturation in the yoke, the magnetic field is usually determined by Eq. (5.1), which from experience is accurate within 1 to 0.5%. On Fig. 5.2 is shown the magnetization curve for 17-orbit

TABLE 5.1 Main Parameters of Electromagnets

Microtron	Number of Orbits	Type of Magnet	Diameter of Pole (mm)	Magnet Gap (mm)	Overall Size (m)	Magnetic Field (kG)	Field Variation (%)	Cooling
Institute for Physical Problems (Ref. 3)	12	Rectangular	700	110	1.5	2	0.5	Water
Physics Institute (Ref. 66)	12	Rectangular	650	110	1.5	1.5	0.5	Air
Institute for Physical Problems (Ref. 27)	17	Enclosed	750	92	0.9	2	0.1	Water
Institute for Physical Problems (Ref. 10)	30	Enclosed	1100	110	5	3	0.2	Water
Large London machine (Ref. 65)	56	Enclosed	2000	100	20	2	0.01	Air
P.S.L. Univ. of Wisconsin (Ref. 100)	34	Rectangular	1370	100	1.8	2.23	0.1	Water

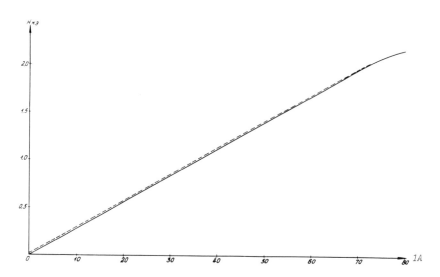

Fig. 5.2 Magnetization curve (dotted line represents the demagnetization curve).

microtron. It is clear that no saturation exists in the operating regions. Consequently, regardless of hysteresis effects and the nonlinearity of the magnetization curve, it is possible to determine the value of the magnetic field from the current to an accuracy of 1%. The field can be measured with greater accuracy (0.01%), by means of a standard IMI-2 nuclear magnetic resonance measuring device.

Field distribution was also measured by using NMR instruments. In those cases where the required accuracy of measurement was greater than that guaranteed by the magnetic power supply stability (approximately 0.1%) we used a differential method of measurement. The difference in the field at two points was determined (see Ref. 24) by the difference in the frequency of proton resonance (for instance in the center of the magnet and at the measured point) according to Fig. 5.3.

For accelerating particles the useful region of the magnetic field is where it is uniform. The requirements for field uniformity depend on the number of orbits N in the microtron and were determined in Section 5 of Chapter IV. For rough evaluation it may be considered that the allowable asymmetry in the nonuniformity is

$$\frac{\Delta H}{H} \sim \frac{1}{N^2} \qquad (5.2)$$

from which it is evident that for large numbers of orbits the field nonuniformity has to be greater.

MICROTRON

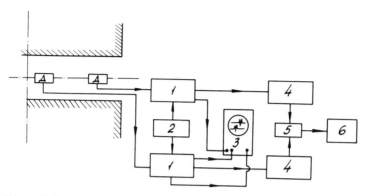

Fig. 5.3 Scheme for measuring magnetic field nonhomogeneity: (0) Field probe NMR; (1) Magnetic field meter; (2) Modulation synchronizer; (3) Double trace oscilloscope; (4) Amplifier; (5) Crystal mixer; (6) Frequency meter.

The uniform region of the magnetic field is increased by shimming the poles (near the edges), which compensate for magnetic flux leakage close to the edges of the magnet. The edge shims are usually designed experimentally; the corresponding field distribution is shown on Fig. 5.4, from which

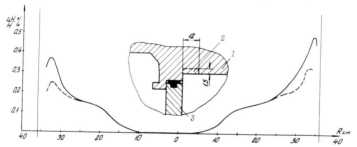

Fig. 5.4 Pole shape and field nonhomogeneity curve: (1) Pole face; (2) Machined circular gap; (3) Chamber. Solid curve - initial field distribution. Dashed curve - field distribution with the circular gap.

it is evident that a uniform field within the tolerance of 0.2% can be obtained in a circular region of diameter $D = D_M - h$, and in the optimum case in a diameter $D = D_M - 0.9 h$ (D_M is the diameter of the magnet poles.

The magnet of our first microtron had poles which were flat all the way to the very edges. In the 17-orbit microtron, to compensate for magnetic resistance caused by finite permeability, the poles were made somewhat convex by approximately 0.17 mm.

In the 30-orbit microtron magnet each pole consisted of two discs, 80 mm thick, separated by a gap of 1.5 mm (see

Fig. 5.5). This gap was used for the placement of shims which compensated for magnetic resistance at the center of the poles.

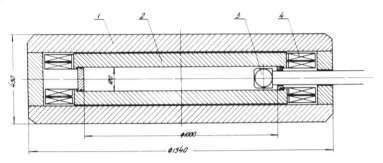

Fig. 5.5 The magnet for the 30-orbit microtron: (1,2) Pole discs; (3) Coils.

From the standpoint of electron dynamics the more serious magnetic field nonuniformity is not axysymmetrical, which is characteristic of axisymmetrical magnets, but field distortions which deviate from symmetry of revolution. The most undesirable is transverse misalignment of the field which leads to orbit drift along the common diameter of the machine. The estimate given in (5.2) is primarily for such field distortion.

Field nonuniformities arise from inaccurate preparation and assembly of the magnet, and from magnetic nonhomogeneity of the material used for the poles of the magnet itself. In the 17-orbit microtron the poles were parallel to within 0.03 mm which, with a gap of 92 mm, should have produced a field misalignment of 0.03%. In fact it was 0.10%.

In the large microtron, magnetic field measurements have shown[24] the presence of an axisymmetric nonuniformity of approximately 0.2% which is strongly dependent on the magnet's former history (Fig. 5.6). Nonuniformity of this type is due

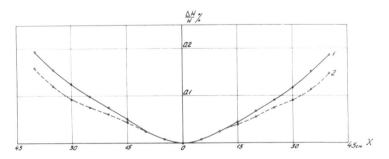

Fig. 5.6 Field nonhomogeneity in the 30-orbit microtron (1) Field established on going up from the zero value; (2) Field established on going down from the maximum value.

to hysteresis and a nonuniformity in the coercive force of the steel in the various sections of the yoke. Besides this, we discovered an asymmetric nonuniformity of the order of 0.03% caused by an inaccurate machining of the flux return path. This nonuniformity was corrected by making a corresponding adjustment of the top pole of the magnet.

The presence of hysteresis was noted in a detailed investigation of the 2-m magnet of the London microtron.[65] The design of this magnet is given in Fig. 5.7. In order to avoid field nonuniformity and its ambiguous dependence on the current a special procedure for switching the magnet was developed.

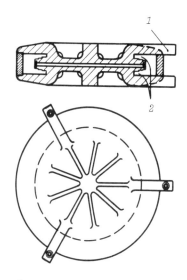

Fig. 5.7 Magnet cross section of the London microtron: (1) Ribs for removing the poles; (2) Rosin for the coils.

In the London microtron the field in the magnet was corrected by a system of concentric coils placed on the poles. This magnet also had a special built-in field nonuniformity close to the cavity which insured beam focusing in the accelerator.

The main problem in our microtrons was the formation of a uniform field, since the beam was focused by the cavity. In order to improve focusing in our large microtron, special coils were used which created a local disturbance in the field close to the cavity thus slightly displacing the whole system of orbits so that the focusing action of the cavity was enhanced.

The magnet coils for the small microtrons were made of copper band 25 × 0.6 mm, wound directly onto the poles; the coil insulation was paper (cable-type ribbon paper laid between the copper band). The coils of our large microtron were wound on a common frame in two sections. In the gap between the magnet and the coils a polyethylene tube which carried water for cooling the coils was placed. The current density was 3 A/mm². However, this low density does not lead to excessively large size of coils and excessive copper weight (in comparison with other accelerators), since the field in a microtron is comparatively small. The coils in the Dubna microtron were made of hollow copper conductors and occupied a larger volume than the system of coils designed by us and used in all the magnets at our laboratory.[37]

The position of the median plane can be regulated by shunting of the magnet coils (Fig. 5.8). Evidently for the

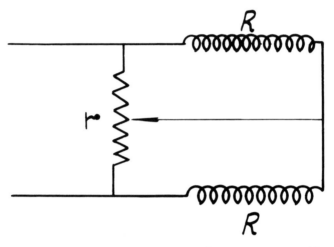

Fig. 5.8 Leads and shunts of the magnet coils: R, Magnet coil; r, Potentiometer.

condition where $R \ll r$, the small changes in the arms of the potentiometer bridge do not appreciably affect the common current but do affect its distribution between the coils. Such a method permits the monitoring of the symmetry of the magnet median plane simultaneously with the position and displacement of the beam.

A considerable nonuniformity of the field was discovered in the Dubna microtron. The geometry of this magnet was similar to the geometry of the magnet in our large microtron; however, during its preparation certain errors and deviations from the design produced a large nonuniformity of the field. For this reason it was necessary to use a very complicated system of corrective coils which made it possible to correct to a high degree of accuracy the position of the median plane of the magnetic field. The current in these coils is tuned with regard to the beam.

In connection with the need for correcting the magnetic field in microtrons, which is especially important where there are large numbers of orbits, S. P. Kapitza proposed a system of coils which would regulate individual motion on each orbit. Such a system can consist of coils placed at the poles (or in the gap between the poles and the magnet), situated along interconnecting regions of corresponding orbits. The leads to the coils should not affect the motion and thus, it is convenient to position them orthogonally to the original coil. Actually, the leads can be directed along the common diameter of the orbits (see Fig. 5.9).

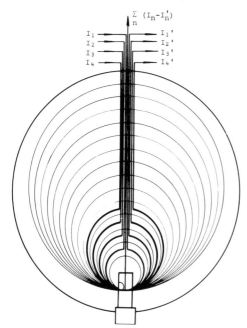

Fig. 5.9 Schematic of the corrective coils.

The current flowing along the contour displaces the field only inside this contour, since its magnetic flux is contained within the magnet. Consequently, the difference in the current in the central coils will determine the displacement of only those orbits which fall between them. By sending different currents through the right-hand and the left-hand sides of the coils (Fig. 5.9), one can displace the orbits along their common diameter. The difference in the current through the coils placed on the top and the bottom poles corrects the position of the median plane of the magnetic field.

It is hoped that this system will insure stable acceleration for a much larger number of orbits than are presently being used in microtrons, and it seems imperative in a cw microtron.

SECTION 3 - THE MICROTRON CAVITY

The cavity is undoubtedly the microtron's main part: by the alternating electromagnetic field, electrons are emitted, captured, accelerated and focused. The microtron cavity in this way embodies functions which in other accelerators are done independently and separately. Primarily because the microtron cavity performs so many functions simultaneously, it has, during the time that accelerators of this type have existed, experienced the largest development.

In the operation of a microtron it is imperative to insure the electrical strength of the cavity in which the field can reach 600 kV/cm. Continuous operation at such a high potential was made possible through a series of developments.

In our experience, cavities have been made from OFHC copper of the type MB. Surface preparation in the cavity is usually limited to mechanical polishing; electropolishing apparently does not increase the resistance for electrical breakdown. After fabrication and polishing, the cavity is usually outgassed in a vacuum at a temperature of 500°C.

At a potential up to 300 kV/cm, vacuum down to 10^{-4} mm Hg does not seem to have a noticeable effect on electrical breakdown. However, at higher fields, which are necessary for acceleration, continuous operation is only possible under conditions of higher vacuum. In our large microtron and in the Dubna microtron, a vacuum of 2×10^{-6} Torr is maintained.

We must emphasize that in flat cavities a potential field of 600 kV/cm does not produce any noticeable field emission; even at 10^6 V/cm the field-emitted current density from copper is approximately 10^{-6} A/cm^2.

For this reason it would seem that breakdown and arcing in an unconditioned cavity have a different origin. Cavity conditioning is accomplished by a gradual increase of the field. Experience on the Dubna microtron has shown that after several tens of hours of continuous operation, the electrical strength of the cavity begins to decrease. The reason for this is still not clear.

Let us note that in a microtron cavity there are no conditions which would lead to the development of high frequency resonance discharge (multipactoring) between the walls. However, there are some indications that there are resonance discharges close to the flight holes in the cavity of the 30-MeV microtron.[28]

The discharges in the cavity undoubtedly come from the presence of films from vacuum pumps utilizing oils* as well as a layer of LaB_6 which is deposited inside the cavity from the continuous evaporation of the emitter. This discharge could also be due to secondary electrons which are knocked out by the beam.

The gap in the toroidal cavity (9 mm long) leads to even higher fields due to its curved surfaces which reach 1.5 to 2×10^6 V/cm. Furthermore, experience has shown that such cavities are capable of operating for long periods of time in conditions of oil-less pumping.[71]

As a rule, we use cavities in which one or both lids were attached to the bodies by screws. All the other parts were brazed; silver brazing of the cavity was accomplished (without soldering flux) in a vacuum at a temperature of close to 800°C

---

*In most modern microtrons Vac-ion elements have led to marked improvement of the vacuum and no oil seems to enter the system any more.

It is also possible to use cavities which are totally brazed. The cavities are cooled by internal channels or by copper tubes soldered onto their body.

The construction of the assembly cavity is shown in Fig. 5.10. In a cavity of this design it is easy to replace the

Fig. 5.10 Cavity of the 17-orbit microtron.

lids and in this way change its geometry. The shape and dimensions of the cavity are determined on the basis of numerical calculations of the modes of acceleration while the shape and dimensions of the holes are determined by the theory of focusing (see Chapter IV).

The coupling between the cavity and the waveguide is accomplished through a hole in the side of the cavity. The dimensions of the connecting opening are usually determined by a cut and try method, measuring the coupling factor by standard measurements of the Q factor and the coefficients of reflection.[26]

The Q factor of assembled cavities is usually between 8000 and 10,000, which is 15-20% less than the calculated value. The Q factor of completely brazed cavities reaches 90% of the calculated value.

Tuning is accomplished by deforming the inner surface of one of the walls, where a circular membrane 0.15 to 0.2 mm thick is machined. The tuning mechanism deforms the membrane and displaces the central part of the wall, thus allowing a tuning of the cavity between the limits of 10-20 MHz. We used two methods for actuating the tuning mechanism; a screw mechanism (Fig. 5.11) and a cam (Fig. 5.12). Both of them appear to operate well, however we find the cam more convenient and thus it is more widely used at present.*

---

*In the latest models direct tuning by a plunger has led to better

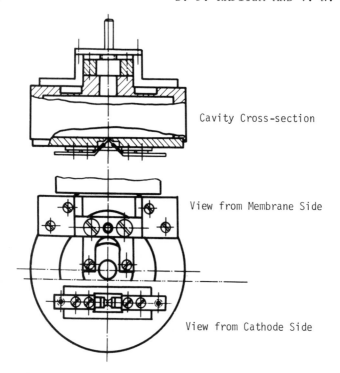

Fig. 5.11 Screw mechanism for the cavity tuning.

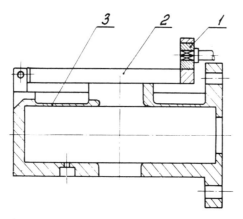

Fig. 5.12 Eccentric mechanism for cavity tuning: (1) Cam; (2) Lever; (3) Membrane.

tuning. Although the tuning range is less than in the case of a membrane, the cavity is more stable in operation.

The cavity inside the chamber is fixed by two methods. In large machines, the cavity is clamped at the end of the waveguide (see Fig. 5.13). Connections for water cooling

Fig. 5.13 A view of the waveguide and cavity (Dubna microtron).

tubes, current conductors, and the tuning mechanism are built in. Consequently, the cavity is a self-contained unit which can be removed from the vacuum chamger through a tube without disassembling the magnet. The cavity can also be adjusted externally.

This design is convenient in large machines where it is difficult to open the chamber due to the large size of the magnet and also due to the induced radioactivity arising from prolonged operation at high energies. It is convenient to assemble and check out the cavity unit separately on a test stand where it is possible to determine its frequency and the coupling coefficient as well as to check the position of the emitter.

In small machines, the cavity is attached directly to the chamber of the accelerator and the waveguide is brazed to the chamber; under the cavity on the inside of the chamber there is a flat access port. Such a design necessitates the opening of the chamber each time the cavity or the emitter has to be replaced, which in small machines does not present any difficulty. The accuracy of cavity placement relative to the median plane of the magnet is 0.1-0.3° of angle and 1-2 mm vertically. Generally speaking, when there is uniform field the microtron does not have a preferential median plane. Thus, the tolerance on vertical placement is only needed to insure the proper position of the beam relative to the extraction tube or the target. The tolerance on the angular deviation of the cavity is determined by the focusing properties of the aperture (see Section 4, Chapter 4).

SECTION 4 - ELECTRON SOURCE

The electron source is placed inside the cavity of the microtron. We have, from the very start, used lanthanum boride as the material for the emitter. This amazing substance thus far has no rivals; it provides a high and stable emission of electrons under a strong alternating field.

Lanthanum boride (LaB_6) is a hard micro-crystalline substance with a density of 4.6 g/cm³, of purple violet color and a melting temperature of 2200°C. The hardness is due to a lattice, similar to that of diamond, consisting of boride atoms bonded by homopolar bonds, while its metal-like conductivity and high emissivity are due to the lanthanum atoms located at the junctions of the lattices. At a temperature of 1600° to 1700° the emitter current density in weak fields reaches 25 A/cm². In strong fields, the emission rises drastically due to the Schotky effect. Special measurements have shown that the density of the emitter current is adequately represented by the Schotky equation

$$i = AT^2 \exp\left(-\frac{\phi - e\sqrt{eE}}{kT}\right), \qquad (5.3)$$

where ϕ = 2.64 eV is the binding energy; A = 73 A/grad²·cm² is the Richardson constant. The dependence i(E) measured by A. B. Vaganov is shown on Fig. 5.14, where it is clear that

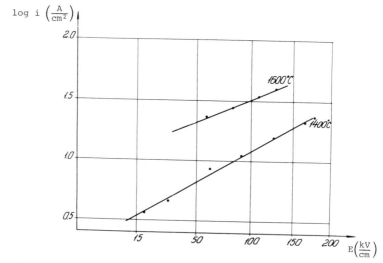

Fig. 5.14 The dependence of emission from lanthanum boride on field and temperature.

there is a pronounced rise in the current density in fields of the order of 100 kV/cm. In the field of the microtron cavity, the emitted current density reaches 100-200 A/cm². It must be emphasized that in contrast to cold emission, the current density under these conditions still depends to a large extent on the temperature.

Lanthanum boride is manufactured in the form of sintered rods which are cut and shaped by electric spark methods.

The electron source is made in the form of a separate unit which is attached to the wall of the cavity and is fixed by means of controlling rods (Fig. 5.15).

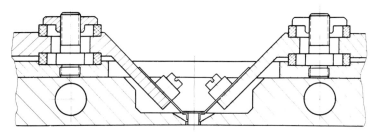

Fig. 5.15 Direct heater cathode.

In our 12- and 17-orbit microtrons the emitter was made in the form of a cube $1.5 \times 1.5 \times 1.5$ mm^3 of lanthanum boride brazed to a tantalum plate 0.2 mm thick. Brazing was accomplished in a vacuum at 2000°C with molybdenum silicate (MoSi$_2$). Heating the emitter was accomplished by an ac current (40-50 A at a frequency of 420 cycles).

The magnetic field of the heater current perturbs the electron motion. In order to avoid this effect the high frequency pulse is synchronized with the moment when the heating current passes through the zero point. If, however, it is evident from the accelerated current that the optimum phase differs from zero, then it is safe to assume that either the emitter or the cavity in the microtron are not positions accurately enough. The emitter is adjusted by special positioning screws. The emitter is usually placed approximately 0.2 to 0.3 mm below the internal surface of the cavity (see Section 6, Chapter III).

The electron source described above is simple and quite dependable, however its lifetime is limited to several tens of hours because of the deterioration of the braze. A more satisfactory method of construction is one in which the lanthanum boride cube is pressed into the tantalum plate. Such emitters* last about a hundred hours at a pulsed current of 1 to 1.5 A.

The Dubna microtron makes use of emitter heated by an electron beam. Emission takes place from a cylinder of lanthanum boride 3 mm in diameter and 5 mm long, held in a tantalum holder. Electrons emitted by a supplementary tungsten spiral filament are accelerated along the magnetic field by a potential of 300-600 V and part of them (30-50%) strike and heat the emitter (Fig. 5.16). The electron source is regulated by a special circuit which stabilizes it and prevents arcing. The emitter current reached 5 A and the lifetime of the electron source was many hundreds of hours.[89]

---

*This technology was developed by V. V. Zhychkov.

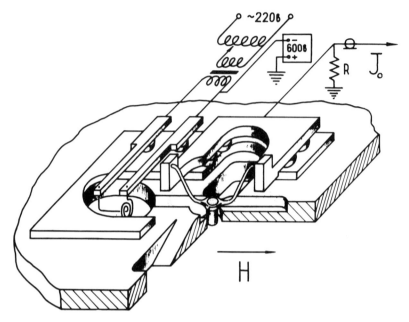

Fig. 5.16 Indirect heater cathode (Dubna microtron).

For operation at a low current mode (up to 1 A) when studying electron optics in an accelerator, it is convenient to use directly heated tungsten or tantalum emitters (see Chapter VII).

Other devices which appear to be of interest as sources of electrons are coaxial electron guns developed by Wernholm[59] for electron injection into toroidal cavities of microtrons (Fig. 5.17). This gun operates at a pulsed voltage of 80 kV

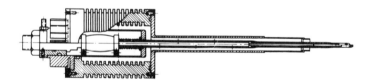

Fig. 5.17 Wernholm's coaxial gun.

and supplies a current of several amperes from a cylindrical lanthanum boride emitter 5 mm in diameter. The emitter is brazed to a straight tantalum plate which is electrically preheated. The gun is placed close to the cavity and can be rotated about its axis in order to optimize its relative position to the cavity apertures. The tube containing the coaxial

MICROTRON   115

gun passes through an opening in the magnet pole. Close to
this opening there are small rings which compensate for the
magnetic field disturbance due to this opening. Further development of this injection method has recently made it possible for O. Wernholm to obtain a current of 175 mA at an
energy of 7 MeV on the microtron of the Royal Technological
Institute in Stockholm.[71]

For brazing the emitter and testing electron sources,
it was found to be very convenient to have a vacuum test rig,
shown in Fig. 5.18. This rig is equipped to have a magnetic
field which allows one to simulate the actual conditions in
the accelerator.

Fig. 5.18 Setup for preparing and testing emitters.

Finally, let us emphasize that the electron source is
extremely important. The quality of the source, the method
of its preparation and the accuracy of its installation to
a large extent determine the successful operation of the
accelerator.

SECTION 5 - BEAM EXTRACTION

Electron extraction from the chamber in a microtron is
relatively simple since the distance between orbits (equal
to λ/π) is quite large. For S-band microtrons this step is
30 mm; the maximum transverse dimension of the electron
bunches along the common diameter of the orbit is equal to
5-6 mm.

Particle extraction is accomplished through a simple
steel tube which shields the magnetic field in the vicinity
of the last orbit. At the entrance, this tube has an internal diameter of 8 mm and a wall thickness of 1.5 to 2 mm.
Gradually its diameter increases to 15-20 mm and the wall
thickness to 5-6 mm. The position of the tube is determined

by the distribution of the orbits and it is usually fixed with the aid of two servomechanisms which displace it in the median plane. Correcting the position of the tube other than along the median plane is usually not needed. In our 17-orbit microtron the beam can be extracted from the 7th to the 16th orbit. In our large microtron, extraction is provided only from the last two orbits. Extraction efficiency is extremely high, being almost equal to 100%. The easiest way to determine the loss of electrons is by measuring the current in the extraction channel, which is usually insulated from the earth.

The channel distorts the magnetic field. This effect can be corrected by steel rods placed close to the exit as was done in our 17-orbit microtron (Fig. 5.19). In the large

Fig. 5.19 Magnetic channel in the 17-orbit microtron.

Dubna microtron the field in the extracting tube was corrected by two plates (Fig. 5.20). In either case, the distortion due to the extraction channel can almost completely be eliminated.

Magnetic channels of a special shape were used in the positron microtron of the Physical Institute of the Academy of Sciences, USSR.[67] Two short tubes were placed inside the chamber such that they changed the position of the last orbit. Field distortion introduced by these tubes was corrected by placing a coil carrying a current proportional to $\cos \phi$ around a magnetized tube. Thus it is possible to completely neutralize the distortion created by the channel in the external space, without creating a field inside the tube itself. The coils were made of copper ribbon. Because of the high current density in the coils a pulsed mode synchronized with the operation of the microtron was used.

The position of the electron bunch in the microtron can also be adjusted by an electric field. By applying a pulsed

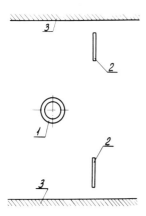

Fig. 5.20 Compensation for field distortion due to the mathetic channel with the aid of two plates: (1) Magnetic tube; (2) Plates; (3) Poles.

voltage of the order of 10 kV on a system of electrodes, it is possible to direct the bunches along one or the other channel and in this way achieve fast switching of the bunches.

The fate of the electrons extracted from the accelerator has little to do with the construction of the extraction channel. One only needs to keep in mind that the monoenergeticity and the small angular spread of the beam simplify its channeling.

The construction of the beam line is determined by the use of the beam. Thus, in the Dubna microtron, the beam line consists of a pair of quadrupole lenses situated between the accelerator and the reactor. These lenses focus the beam on the target inside the active zone of the fast neutron assembly at a distance of 10 m where the diameter of the beam on the target is 5-8 mm (see Ref. 89). More complicated beam lines are used where the particles are injected into other accelerators. In the Lebedev Physics Institute, for instance, the beam line consists of four elements providing for a system of parallel translation. In the 1.2-GeV strong-focusing synchrotron at Lund (Sweden) the injection beam line consists of two systems for parallel translation and two triplets of quadrupole lenses; the orbit plane of the 5-MeV microtron is situated vertically since in such a position its vertical and horizontal emittances are more compatible with the admittance of the synchrotron.

SECTION 6 - THE ACCELERATOR CHAMBER AND THE VACUUM SYSTEM

It is practical to make the vacuum chamber of the microtron part of the magnet. In that case, the poles of the magnet act as the flat walls of the vacuum chamber and the cylindrical nonmagnetic ring as its side walls. This principle is used in most microtrons. The side walls of the chambers in our microtrons were made of brass, while in the Dubna microtron they were made of stainless steel.

The seals for the ring and the poles are made of rubber, which is simple and dependable. For instance, the same seal on our small microtron was used for four years with practically daily use.

The cylindrical wall of the chamber contains feedthroughs for cables and mechanical devices, the construction of which follows standard procedures of vacuum technology. Inside the chamber, along the common diameter of the orbit, there is a probe which is inserted through the side of the wall. It is convenient to use this probe for placing targets and other devices for measuring the size and shape of the beam current pulse on all the orbits. The probe and the target itself are isolated from the chamber. The target and probe during current observation pick up interference largely associated with the high frequency oscillations in the chamber that are detected by secondary electrons as well as the presence of discharge in the chamber.* In order to avoid these spurious signals it is best to use a shielded target (in the form of a Faraday cup) or by using a target which is isolated from the probe. However, for accelerated currents exceeding 1 mA it is quite possible to measure the average current and observe pulse shapes by using a target which is not insulated from the probe. If there is excessive heating of the target (more than 50 to 100 W), it is necessary to water-cool it.

Vacuum pumping in the microtron is accomplished by means of diffusion pumps. In our first microtron we used two pumps with a capacity of 100 and 500 liters/sec which pumped the chamber and the waveguide. Such large systems insure fast and dependable pump-downs of the volume, an expedient which is especially useful for frequent dismantling of the chamber during experiments with the accelerator. Thus, with hot pumps, operating vacuum was reached within 10 to 20 minutes after closing the system. Our 17-orbit microtron initially had a 100-liter pump with semiconductor trap, which seemed to be entirely adequate for prolonged use of the machine. To speed up the pump-down, however, we installed a 500-liter pump with a nitrogen trap. The above-mentioned machines operated quite well with pressures starting at 5×10^{-5} Torr. Vacuum measurements were made with standard meters such as the magnetic Penning discharge or the ion-gauge types.

The vacuum system for our large microtron consists of two 500-liter diffusion pumps which evacuated the chamber and the waveguide. The operating vacuum in this microtron was slightly higher equaling 2 to 3×10^{-6} Torr.

We must conclude, however, that in the future it would be necessary to use oil-less pumping on microtrons, especially in large ones. In the Dubna microtron, for instance, the vacuum system was improved and is now accomplished by means of three electrodischarge systems. The main vacuum seals in it are rubber; however, in place of the Wilson-type seal, bellows are used at the present time.

---

*The magnitude of these interference signals depends on the position of the probe; apparently, in this case, the probe tunes the chamber in resonance with the accelerating field or one of its harmonics.

MICROTRON

SECTION 7 - HIGH FREQUENCY GENERATORS AND AMPLIFIERS AND DESIGN OF THE WAVEGUIDE SYSTEM

The high frequency system of a microtron consists of the generator, the waveguide, and the cavity.

The transmission of power must not only be efficient, but must also be stable from the standpoint of small shifts in the frequency of the generator or the frequency of the cavity loaded by the electron beam. All the microtrons constructed thus far have operated in the pulsed mode with considerable amounts of power, measured in hundreds or thousands of kilowatts, while the pulse length is approximately 1-5 μsec at a duty cycle of 10^{-3}. Consequently, at the start of each pulse the system must enter the operating mode without any instability and without operating on a parasitic frequency. For this reason, the stability of the operating system in a pulsed accelerator is especially important.

In Table 5.2 are given some characteristics of power generators in the 10-cm waveband, which can be used for microtrons.

TABLE 5.2 Certain Characteristics of High Frequency Amplifiers and Generators

Device	Power (MW)	Efficiency (%)	Anode Voltage (kV)	Amplif. (dB)	Freq. Line (MHz)	Remarks
Klystron	20	35-40	250	30-40	100	Entrance and exit completely independent
Magnetron Amplifier	5	65-70	50	10	100-50	Entrance and exit not independent
Magnetron	2	50-55	35-50	-	3	Frequency shift at SWR = 1.5
Nigotron	0.170	50	30	-	-	Continuous wave magnetron (Ref. 30)

In microtrons, the most common generators are powerful pulsed magnetrons. Developed for radar, these generators can develope power up to several megawatts. Their use in microtrons is entirely feasible if they are matched to the cavity by ferrite isolators or parallel dissipative loads. The frequency of the microtron, like any other self-excited generator, depends on the load. The frequency pulling, which in radar has a detrimental effect, with a microtron cavity acts as a stabilizing factor on the operatint system as a whole.

If a nontunable magnetron is used, then tuning of the system is accomplished by varying the frequency of the cavity. Where tuned magnetrons are used, then the construction of the cavity is simplified since there is no need for tuning elements. On the whole, it can be said that magnetrons are simple and cheap microwave generators, especially in the case of power requirements from 2-5 MW. The efficiency of a magnetron is usually higher than that of a klystron, approaching between 50 and 55%; besides, the magnetron operates at a substantially lower anode voltage than klystrons, which is also an advantage.

Within the last several years, magnetron amplifiers or amplitrons have been developed. These microwave devices have an even higher efficiency (up to 70%) but a rather low coefficient of amplification (about 10 dB). Thus far, these systems have not been used in microtrons, however their use would seem to be reasonable especially for high powered microtrons.

Powerful klystrons were developed for use in linear accelerators. At present, klystrons are used in two microtrons: the injector into the Lund synchrotron[59] and in the Dubna microtron.

Klystrons are obviously more complex and more expensive than magnetrons. Their use can be justified in unique machines where the efficiency, simplicity and low cost of the system are of secondary importance and can be sacrificed for the sake of the large power and the high coefficients of amplification which are readily available with klystrons.

With pulsed magnetrons, the system shown on Fig. 5.21 can be used. A relatively simple system (Fig. 5.21a), it provides a parallel load placed at a distance of a quarter wavelength + $n\lambda$ from the cavity. When the cavity is not tuned or at the start of a pulse when the power is reflected from the cavity, the load decreases the amplitude of the reflected wave. When the cavity is in tune, the load absorbs only a small portion of the power. The portion of the power which is absorbed by the load (let us designate this value by x) depends on the limiting VSWR allowable for the magnetron (Fig. 5.22).

This method (along with other methods which do not use ferrites) was investigated by Reich,[53] Stepanchuk,[75] and also by Milovanov[78] and Kantoroy.[69] Some of these circuits transfer 50-70% of the magnetron power to the cavity.

The magnetron is sensitive to the phase of the reflected wave. Large values of x are obtained with the so-called sink phases which correspond to smaller power generation and conversely smaller values of x.

Instead of parallel shunting, it is possible to use an attenuator between the magnetron and the cavity. However, such a design leads to lower efficiency and stability and thus, has not been used.

It was experimentally demonstrated by Stepanchuk[75] that it is possible to operate a cavity directly from the magnetron if the reflected wave comes back in the appropriate phase. This method requires thorough control of the phase and does not allow for any noticeable detuning in the frequency of the cavity and the magnetron. Power transmission

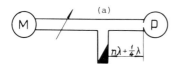
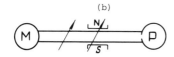
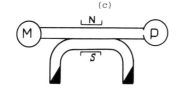
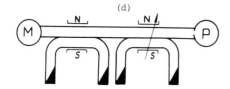

Fig. 5.21 Method of switching the magnetron: M, Magnetron; P, cavity.

in this method is high; however, thus far it has not been possible to operate a magnetron at nominal power mode and with high efficiency.

The compatibility of the cavity with the magnetron is considerably simplified with the use of unidirectional elements such as ferrite isolators and circulators (Figs. 5.21b and c). With this method it is possible to transfer 90% of the power to the cavity when operating the magnetron at high power. Frequency pulling is less than in the previously discussed cases and normally is 2-3 MHz. This is adequate for stabilizing the system. For large power transmission this method is sensitive to the matching of the cavity to the line since the presence of isolators does not allow complete decoupling.

This method of using ferrite isolators has found wide acceptance in microtrons and, in low power machines it seems to be the best.

A more complicated method utilizing circulators is used in our large microtron and also in the Dubna microtron.[28]

In the Dubna microtron, Matora and Kharyuzov[73] proposed and adopted a method which uses two circulators (Fig. 4.21d).

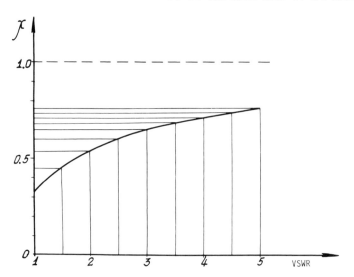

Fig. 5.22 Dependence of the efficiency on V_{SWR}.

The first circulator operates under constant magnetization of the ferrite and accomplishes practically total decoupling of the magnetron from the load. The second circulator, as the magnetization of the ferrite is changed, acts as a regulating power divider. In this method the magnetron operates in an optimum and unchanging mode and its frequency is practically independent of the load. Thus, there is no frequency pulling of the generator with the result that the tuning of the magnetron and the accelerating cavity have to be held within tighter tolerances.

In principle, it is possible to use powerful microwave amplifiers in two types of systems. First of all, these amplifiers can be used as an output load. This method was used in the Swedish microtron injector[59] with a three-resonator Varian klystron. The 1-MW klystron is excited by a pulsed triode, the frequency of which is tuned to the frequency of the cavity. A permanent magnet ferrite isolator is placed between the klystron and the cavity for protecting the exit window of the klystron. This method required that the frequency of the exciter and the cavity be mutually stable to a high degree of accuracy ($\Delta f/f < 3\text{-}4 \times 10^{-5}$).

A similar method using a magnetron amplifier also seems to be possible. However, in this case because of the smaller amplification coefficient of the amplitron, the exciting power will be substantially large.

It is somewhat interesting to utilize a high power microwave amplifier in a self-excited mode where the feedback loop includes the cavity of the accelerator. This method was proposed by S. P. Kapitza and was used in the Dubna microtron by R. V. Kharyuzov, S. I. Golubev, and E. M. Matora.

To make use of a klystron in a self-excited mode it is necessary to divert a small amount of power (10^{-3} to 10^{-4})

MICROTRON   123

from the cavity, and apply it at the input of the amplifier.
In view of the insignificance of this power in the Dubna
microtron, a coaxial line feedback coupling was used connecting the cavity through a coupling loop. In this method the
only low power element in the system that needs adjustment
is the phase shifter in the feedback circuit. For controlling the amplitude, it is reasonable to include an attenuator (Fig. 5.23) in this circuit. The connection between the

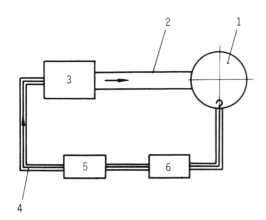

Fig. 5.23 Scheme for switching on the klystron with the
use of a feedback loop through the cavity: (1) Cavity;
(2) Waveguide; (3) Klystron; (4) Feedback loop; (5) Attenuator; (6) Phase shifter.

output cavity of the klystron and the accelerator cavity can
be made either directly or through a ferrite isolator (the
latter is important for protecting the klystron window in
case of breakdown).

The passbands in powerful klystrons are such that the
cavity does not need tuning elements. This method also in-
troduces some protection against a breakdown in the accelerating cavity since the feedback is discontinued when a discharge
occurs.

In this method it is also possible to use a magnetron
amplifier; the cavity also must have an additional coupling
element. Because of the small coefficient of amplification
the feedback coupling must handle no less than 10% of the
power applied to the cavity. Thus, the feedback coupling loop
must be a waveguide, especially since the input coupling for
most magnetron amplifiers, unlike the klystron, is also a
waveguide.

The waveguide feedback coupling chain must have a ferrite
isolator and/or a phase transformer which protects the amplitron from the parasitic generation caused by reflections from
the cavity (the amplitron, in contrast to the klystron, is
not a unidirectional device). The high efficiency, the low

voltage, the simplicity, and the dependability of the magnetron amplifier make it ideal for this kind of system.

The conditions of excitation in systems utilizing klystrons and amplitrons are similar and can be easily deduced. Let us only note that the klystron can be approximated as a linear amplifier; its amplification factor which is constant for small signals and power, slowly decreases as the signal level increases. Magnetron amplifiers are definitely nonlinear devices: for small signals their amplification factor can be much larger than for large signals where they rapidly reach saturation.

In these systems the generated frequency is determined by the frequency of the cavity, which is especially important in powerful accelerators where the beam pulling can cause a noticeable shift in the frequency of the cavity. This advantage of these systems along with the others mentioned leads us to conclude that the use of powerful amplifiers is an important step in the development of microwave systems for microtrons; the use of amplifiers will both simplify the accelerator and make them more efficient.

SECTION 8 - THE MICROWAVE SYSTEM OF THE MICROTRON

The microwave generators in our microtrons were standard nontunable pulsed S-band magnetrons. Pulse length in the small microtrons was 1-3 µsec at a nominal operating frequency of 427 Hz giving a duty cycle of about 1/800.

The modulator in the lesser machines consisted of a pulse-forming network and transformer which matches the line to the magnetron. The line network is charged through a resonance choke. Switching is accomplished by hydrogen thyratrons. The power of the magnetron (from 50-100% of nominal capacity) varies depending on the pulse voltage. The modulator in the 30-orbit microtron was a hard tube pulse forming system with a storage capacity charged to the operating potential of the magnetron. This system allows the repetition rate and the pulse length to be regulated within wide limits; the rate can be varied from several hertz to 1000 Hz and the pulse length from 1-5 µsec.

Pulse repetition rate is controlled by an external generator which is synchronized with the frequency of the primary power network (400 cycles), its harmonics and subharmonics: 300, 400, 200, 100, 50, 25 or 12½ Hz. There is a phase transformer in the synchronization system which phases the trigger pulse. This is convenient in the case where directly heated emitters are used in which the field of the heating current (approximately 30 A) displaces the position of the orbit. The Dubna microtron was synchronized with a rotating uranium reactor insert, and normally operated at a frequency of 50 Hz.

The microwave circuit of our first microtron is shown in Fig. 1.9. It consists of a phase transformer, a vacuum window and a waveguide branch with a matched water load. This load dissipated approximately 40-50% of the power. The system is simple and insures adequate matching of the magnetron to the cavity.

MICROTRON   125

A more advanced method was used in the 17-orbit microtron. A ferrite isolator was used between the magnetron and

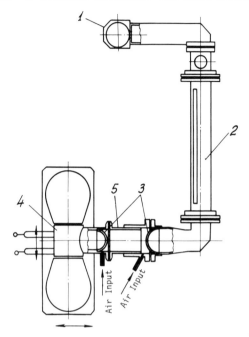

Fig. 5.24 Waveguide for the 17-orbit microtron: (1) Cavity; (2) Ferrite; (3) Vacuum glass; (4) Magnetron; (5) Phase shifter.

the cavity. In this case the losses in the guide were approximately 15-25%. The magnetic field in the ferrite is usually larger than the field which corresponds to the largest decoupling; in this way the losses in the ferrite are reduced and the operating stability of the magnetron is increased. In order to tune this system, it is necessary to regulate the electrical length of the waveguide; this is accomplished by displacing the position of the magnetron, by means of a sliding contact (a trombone) situated in the waveguide between the magnetron exit flange and the vacuum window (at this point in the tract, a circular waveguide is used).

In all these systems, the vacuum window is situated at the beginning of the waveguide immediately after the exit window of the magnetron where it is not likely to have large standing wave ratio.

In the waveguide of the large microtron and of the Dubna microtron there were two circulators separated from the magnetron and the cavity by vacuum-tight windows. In the part of the waveguide where ferrite elements were placed it was possible to fill the system with high pressure nitrogen when large powers are dissipated.

In small accelerators it is desirable to use ferrite elements separated from the accelerator vacuum system since on one side, the properties of ferrites are not very compatible with vacuum, and the gas which fills the waveguide improves the cooling of the ferrite.

Suitably placed coupling loops are useful for measuring the high frequency signals. The signals from these loops allow observation of the shape of the pulse or are used for measuring the frequency by a wave meter. A convenient but rather bulky element of this system is a waveguide directional coupler situated immediately in front of the cavity. Together with an oscilloscope, it permits monitoring the incident and the reflected waves and these oscillograms provide complete information on the coupling of the cavity to the tract, the loading of the cavity by the beam, and of any other transient processes occurring during accelerator operation.

SECTION 9 - POWER SUPPLY AND CONTROL SYSTEMS

At the Physical Laboratory electrical supply comes from three networks. Vacuum and circulating pumps, blowers and other systems which do not require a high stability in the voltage use an unstabilized 50-Hz main. Next come the supplies at a frequency of 50 Hz which are stabilized by ferroresonance stabilizers to ± 0.5%. Finally, the magnetron modulating networks, the microtron and ferrite magnet rectifiers, heating circuits for the emitters, are supplied by a stabilized source ± 0.2% provided by a motor generator of the type VPL-30, at a frequency of 427 Hz. This frequency of 427 Hz is convenient for several reasons. First of all, magnetic amplifiers, transformers and filters are much more compact and more simple in their construction at a frequency of 427 Hz. Besides stability, this system also has the advantage that it is asynchronous to 50 Hz, and the 427 Hz make the various kinds of disturbances easier to identify. We should also mention that most conventional laboratory apparatus can be presently operated at a frequency of 427 Hz.

The current in the magnet correcting coils and the modulators was controlled and stabilized by a system of magnetic amplifiers.

In these accelerators we make wide use of water cooling of magnetrons, cavities, ferrites, magnetic coils, etc. Cooling is usually done with distilled water circulated in a closed system and cooled in turn by the city water main.

For power measurement the water is directed first through tubes which cool the cavity and the ferrite and then through a calorimeter (Fig. 5.25). The temperature difference before and after the cavity, the ferrites and the calorimeter is measured by identical copper constantin thermocouples and the signals are automatically recorded on a multichanneled potentiometer of the type EPP-09. The calorimeter is a coaxial heated unit capable of dissipating 1 kW; this power is measured by an electrodynamic watt-meter. This method allows one to calibrate the potentiometer and measure the power loss in the ferrite, the cavity or the target simultaneously without the need of measuring the water flow and calibrating the thermocouples.

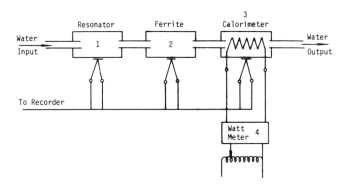

Fig. 5.25 Setup for measuring power: (1) Cavity; (2) Ferrite; (3) Calibrator; (4) Wattmeter.

The shapes of the high frequency pulse, the emitter current and the current on the target were observed on oscilloscopes. The remaining parameters were controlled by gauge systems.

For determining the mode of acceleration which corresponds to the required energy, it is convenient to have a nomogram such as the one shown on Fig. 5.26. This nomogram allows one to find the number of the orbit n and the parameter Ω (or the current of the magnet) for which the beam will have the total energy U. On the abscissa the parameter Ω which is proportional to the magnetic field and the current in the coils of the magnet is given, while on the ordinate the kinetic energy in mega-electron volts is shown. The inclination of the curves depends on the orbit number.

Accurate determination of the total energy on the nth orbit is accomplished by measuring the frequency of the magnetron, f, and the magnetic field, H, which are best determined by the frequency of nuclear magnetic resonance of protons f_p. We have

$$U(MeV) = 0.3 \; H \; (kOe) \; R(cm) = \frac{4500}{\pi \gamma} \frac{f_p}{f} (n + 1) , \qquad (5.4)$$

where the energy U is expressed through the ratio of two frequencies and the gyromagnetic ratio for protons (γ = 4.26 MHz/kOe). The actual accuracy of energy determination depends on the uniformity of the magnetic field (usually the nonuniformity is less than 0.1%) and the beam energy spread associated with phase motion. Energy spread is usually ± (5-20) Ω keV, and is determined by the phase motion which, for n > 3, is practically independent of the number of orbits. Within the phase oscillation nodes, energy scattering is less and is approximately 5-8 keV.[38]

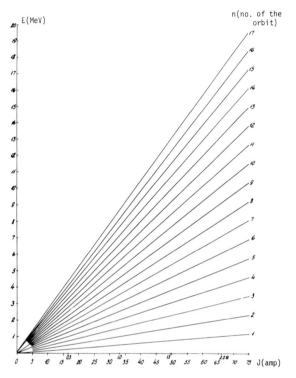

Fig. 5.26   Nomogram for determining energy.

For accurate work the magnetic field is measured by nuclear magnetic resonance, with a standard magnetic field measuring probe. The wavelength of the magnetron was measured by a wave meter VST or by a heterodyne frequency meter.

The control panel for the 17-orbit microtron is shown on Fig. 5.27.

During tune-up and preparation it is convenient to observe the pulse in the cavity from a small inductive loop within the cavity along the emitter current and the current in the beam. The generated signal makes it possible to monitor the frequency and evaluate the amplitude stability of the oscillations in the cavity. Detection is best accomplished by high frequency vacuum diodes.

MICROTRON   129

Fig. 5.27 Control panel for the 17-orbit microtron.

A plan of the building in which the 17-orbit microtron is housed is shown on Fig. 5.28. This plan, which does not require any explanation, has proved to be convenient for running the machine and conducting various experiments. The building is ventilated by fans together with an air conditioner of 10 kW capacity. The power needed by the accelerator is 15 kW, of which 10 kW are required by the magnetron modulator; the power of the beam in this microtron is 0.5 kW.

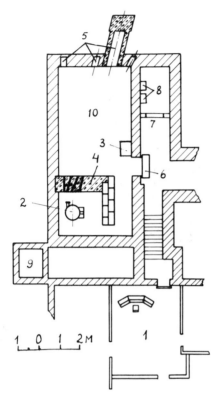

Fig. 5.28 Building plan where the 17-orbit microtron is located: (1) Control room; (2) Microtron; (3) Modulator; (4) Shielding; (5) Radiation traps; (6) Door; (7) Ventilation room; (8) Heat exchangers; (9) Ventilation exit duct; (10) Experimental hall.

CHAPTER VI

PHYSICAL CHARACTERISTICS OF THE MICROTRON

SECTION 1 - THE POWER NEEDED BY THE CAVITY, AND ITS MATCHING TO THE WAVEGUIDE

In analyzing the main physical characteristics of the microtron we will begin with its energy, and will devote Sections 1 and 2 to it. The choice of frequency bands of the microtron and similar problems will be discussed in Sections 3 and 4. High current effects in the microtron will be treated in Sections 5 and 6.

The power dissipated in the walls of the cavity can be calculated by integrating the alternating magnetic field H on the inside surfaces of the cavity; this power is equal to (see, for instance, Refs. 56 and 98):

$$P_r = \frac{c^2}{32\pi^2 \sigma d} \oint |\underset{\sim}{H}|^2 \, dS \, , \qquad (6.1)$$

where σ is the conductivity of the metal, $d = \frac{1}{2\pi}\sqrt{\frac{\lambda c}{\sigma}}$ is the skin depth. By using the expression for the E_{010} mode in a cylindrical cavity and integrating, we obtain

$$P_r = \frac{\pi}{4} \nu_{01} J_1^2 (\nu_{01}) \frac{d}{\lambda} (\nu_{01} + \ell) \varepsilon^2 \Omega^2 P_0 \, , \qquad (6.2)$$

where $\nu_{01} = 2.405$ is the first root of the equation $J_0(\nu) = 0$,

$$P_0 = \frac{m^2 c^5}{e^2} I_0 V_0 = 8700 \text{ MW} \qquad (6.3)$$

is the characteristic electron power, of fundamental importance for accelerators and microwave devices.

In classical electrodynamics[33] the value P_0 can be considered as the product of the voltage

$$V_0 = \frac{mc^2}{e} = 511 \text{ kV} \, , \qquad (6.4)$$

which corresponds to the rest energy of the electron and the characteristic current specific for the electron,

$$I_0 = \frac{mc^3}{e} = 17000 \text{ A} \qquad (6.5)$$

the current I_0 can be envisaged as arising from the motion of relativistic electrons following each other at a distance of r_0, where $r_0 = e^2/mc^2$ is the classic electron radius; in other words I_0 is the instantaneous current carried by the electron.

The values P_0, V_0 and I_0 are the true physical units for various electron machines and thus, they always appear in the introduction of dimensionless variables and, in essence, are convenient to use for calculating and evaluating power and current in a microtron.

The Q factor of the cavity at the frequency of the E_{010} mode is determined by the expression

$$Q = \frac{\nu_{01}\ell}{kd(\nu_{01} + \ell)} \qquad \left(kd = \frac{2\pi d}{\lambda} = \sqrt{\frac{c}{\lambda\sigma}}\right) \quad . \tag{6.6}$$

From Eqs. (6.2) and (6.5), it is easy to obtain an expression for the reactive power $\tilde{P}_r = QP$:

$$\tilde{P}_r = \frac{\nu_{01}^2}{8} J_1^2(\nu_{01}) \; \varepsilon^2 \Omega^2 P_0 = 0.2 \; \varepsilon^2 \Omega^2 P_0 \quad . \tag{6.7}$$

$\tilde{P}_r$ is of the order of P_0 and depends only on the relative potential of the field in the cavity and its length; the value of $\tilde{P}_r$ represents the energy stored in the cavity. Let us note that the displacement current through the cavity is determined by an analogous formula

$$\tilde{I} = 0.2 \; \varepsilon \Omega I_0 \tag{6.8}$$

and to an order of magnitude is equal to I_0.

Since ℓ depends only on ε [see Eq. (3.18)], then the power loss in the cavity depends only on ε and Ω (or ℓ and Ω),

$$P_r = 1.02 \; \frac{d}{\lambda} \left(1.2 + \arcsin \frac{1}{1.88\varepsilon}\right) \varepsilon^2 \Omega^2 P_0 \quad . \tag{6.9}$$

The dependence of the relationship $P_r \lambda / P_0 d$ on Ω, ε and ℓ is shown on Fig. 6.1. For copper at room temperature $\sigma = 5.7 \times 10^5 \; \Omega^{-1} \; cm^{-1}$ and thus, for $\lambda = 10$ cm we have $d = 1.2 \times 10^{-4}$ cm and $P_0 d/\lambda \simeq 100$ kW; this simplifies the use of Fig. 6.1.

An expression analogous to (6.9) for rectangular cavities with sides a, b, and ℓ has the form

$$P_r = \frac{\pi^2}{8} \frac{d}{\lambda} \left[\frac{a^2 + b^2}{2ab} + \frac{(a^3 + b^3)\ell}{a^2 b^2}\right] \varepsilon^2 \Omega^2 P_0 \quad . \tag{6.10}$$

Power loss in a rectangular cavity is approximately equal to that lost in a cylindrical (circular) cavity of the same thickness and comparable transverse dimensions (for the same frequency and $\ell = 1$ the losses in a square cavity are 7% larger than in a circular one). Since the vertical dimension of a rectangular cavity can be reduced by enlarging its transverse dimension (keeping to the same frequency) it is possible to decrease the gap in the magnet. For this reason in certain

MICROTRON 133

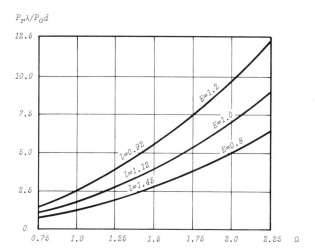

Fig. 6.1 The dependence of ohmic power losses on various parameters.

cases it is advisable to use rectangular cavities, especially if the calculations of the modes of acceleration indicate these advantages.

For a given wavelength, λ, aside from the parameter Ω, the losses in a cavity are mostly affected by the thickness. From the structure of Eq. (6.9) it can be seen that, for a certain thickness, at a given energy gain per revolution the losses will be minimal. On Fig. 6.2 the dependence of power

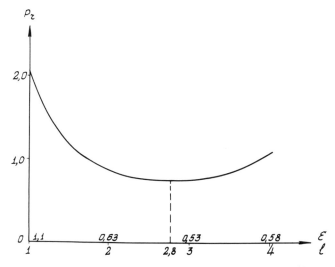

Fig. 6.2 The dependence of power on the cavity thickness.

loss on the thickness of the cavity is shown. The curve indeed has a minimum point--this is a result of the fact that beginning with a certain optimum thickness, the effectiveness of the cavity decreases with increase in the transit angle. For $\lambda = 10$ cm the optimum thickness is equal to 44 mm; cavities in microtrons are usually thinner.

The coupling coefficient between the cavity and the waveguide is equal to

$$\beta = Q/Q_H - 1 \quad , \tag{6.11}$$

where Q_H is the loaded Q factor. Calculation of the aperture dimensions for a given coupling coefficient is possible in principle but not advisable. For this reason, it is common to measure the coupling coefficient (see Ref. 26) by measuring the coefficient of reflection R, related to the coefficient β by

$$R = \left| \frac{1 - \beta}{1 + \beta} \right| \tag{6.12}$$

and the Q factor

$$Q = (1 + \beta) Q_H .$$

For $\beta > 1$ the beam loaded cavity will be matched to the waveguide (no reflected wave) if the total power P applied to the cavity is equal to

$$P = \beta P_r \tag{6.13}$$

and the power utilized by the electrons in the cavity is equal to

$$P_e = (\beta - 1) P_r . \tag{6.14}$$

Without electron loading this cavity will have a coefficient of reflection (6.12).

The connection between the powers P, P_e, P_r and the generator power determine the efficiency of the accelerator to which the following section is devoted.

SECTION 2 - MICROTRON EFFICIENCY

The efficiency of the microtron depends on the high frequency power distribution. The value of the accelerator efficiency itself is relative, and is not so much the economic characteristic of the machine as much as the measure of our ability to optimize the power available and a measure of the progress which allows further improvement of the accelerator.

Actually, the efficiency of the magnetron is usually between 50% to 60%, and the efficiency of a modulator of average power is equal to 65% to 75%, and so, if the efficiency of the accelerating cavity is 30% (i.e., 30% of the power required by the cavity is transferred to the beam) then the relative energy losses in the cavity are not so large. It is

MICROTRON

worth noting that the energy expended on pumps, blowers and coolers is comparable to the energy expended on accelerating the beam of particles and often exceeds the latter. It is also worth remarking that the cost of the microwave generators often exceeds the cost of the energy needed to operate them over their expected lifetime.

Let us first of all consider energy losses in the cavity. If the cavity is matched to the waveguide then the power needed to excite it is equal to (6.13). However, in a pulsed mode of operation, acceleration takes place only when the cavity is fully excited. From the theory of phase motion it follows that acceleration in a microtron is possible only under the condition

$$V > \frac{\cos 32°}{\cos 20°} V_s = 0.9 V_s ,$$

where V is the amplitude of the potential in the cavity, V_s the stationary value which corresponds to the optimum equilibrium phase $\phi_s = 20°$. If at $t = 0$ the cavity receives a rectangular pulse, then the dependence of V on t for $t > 0$ is approximately determined by the expression

$$V = V_s(1 - e^{-t/\tau}) , \qquad (6.15)$$

where τ is the time needed to establish oscillations in a cavity and is equal to

$$\tau = \frac{Q_H}{\pi f} = \frac{2Q_H}{\omega} . \qquad (6.16)$$

If $t = 2\tau$, then $V = 0.87 V_s$, so that in practice, the beam of accelerated electrons appears when $t > 2\tau$.

In the cavity of our 17-orbit microtron $Q_H = 3000$ and $2\tau = 0.7$ μsec. A typical oscilloscope trace of the current shows that the length of a pulse of accelerated beam is equal to 2.4 μsec if the length of the high frequency pulse is equal to 3.1 μsec; this is in good agreement with the estimate. At the end of the pulse the acceleration stops abruptly in less than 0.25τ (after 0.12 μsec in the case mentioned). These estimates are not very accurate since Eqs. (6.15) and (6.16) do not take into account nonlinear effects of the beam loading of the cavity and also processes associated with energy reflection from the cavity as it is not perfectly matched to the waveguide in the absence of a beam.

Thus, for a pulse microtron the important losses are those which occur during the buildup of oscillations. The relative magnitude of these losses is greater for a shorter pulse; as the pulse is lengthened they decrease and in a continuous duty microtron they disappear altogether. In the considered case, the efficiency due to the pulse duration is

$$\eta_T = \frac{2.4}{3.1} = 0.68 ,$$

i.e., these transient losses in a microtron are quite substantial.

It must be pointed out that the strong dependence of the accelerated beam on the amplitude of the voltage causes the accelerated electron pulse to have a rather steep risetime.

Because the current pulse is shorter than the high frequency pulse, it is possible to make short pulses of accelerated beam with relatively long field pulses. Thus, on the 17-orbit microtron, when the length of the accelerating field pulse was shortened to 0.6 μsec, it was possible to obtain sharp current pulses lasting approximately 0.12 to 0.15 μsec; the repetition rate was 2000 Hz.

The main losses are caused by the process of particle capture into the acceleration mode and their subsequent loss. During particle injection only a small portion of electrons possess the proper phase and energy.

An important characteristic of the mode of acceleration is the value ϕ, which is the width of the phase region of capture. The dependence of ϕ on the various accelerator parameters was studied in Section 6 of Chapter III. However, the efficiency of the accelerator is determined not so much by the value of ϕ, as by the coefficient of capture

$$K = J_N/J_0 , \qquad (6.17)$$

where J_N is the current of the accelerated particles (at the last, N^{th} orbit) and J_0 is the total emitted current. The value K depends not only on ϕ but also on a series of other factors which cause particle losses. First, particles to be lost are those which have too large an amplitude of vertical oscillations. Furthermore, the action of various perturbations lead to the excitation of phase oscillations and thus bring about further particle losses.

As it is difficult to calculate the coefficient of capture, it is usually determined experimentally. In the first type of acceleration

$$K = \frac{1}{20} \text{ to } \frac{1}{30} .$$

In the second type of acceleration the region of capture is wider, but vertical focusing is weaker, and thus the coefficient of capture is essentially the same.

If we assume (this assumption is usually justified for current up to 50 MA) that the condition of current drop-off along the orbits does not depend on the current, and if we take into account that the coefficient of capture K is considerably smaller than unity, then the power of the beam of accelerated electrons P_N and the power $P_e - P_N$ needed to accelerate the additional electrons which do not reach the last orbit, can be presented in the form

$$P_N = J_N \, N \, \frac{\Delta U}{e} , \qquad P_e - P_N = (J_0 - J_N) \, \frac{\bar{W}}{e} = \frac{J_N \bar{W}}{Ke} , \qquad (6.18)$$

where ΔU is the energy gain per revolution, $\bar{W}$ is the average kinetic energy of the electrons which are lost during the acceleration process. By using these expressions we can present the efficiency of the mode of acceleration η_N and the efficiency of the accelerator η in the following way

$$\eta_N = \frac{P_N}{P_e} = 1 \bigg/ \left(1 + \frac{\bar{W}}{KN\Delta U}\right) , \qquad (6.19)$$

$$\eta = \frac{P_N}{P_S} = \eta_N \chi \left(1 - \frac{P_r}{P}\right) , \qquad (6.20)$$

where P_S is the power of the high frequency generator; $P = P_r + P_e$ is the total power required by the cavity; $\chi = P/P_S$ is the portion of the power applied to the cavity.

From Eqs. (6.19) and (6.20) a series of conclusions can be made. One of the most obvious is the fact that as the power of the generator P_S is increased, the portion expended as ohmic losses becomes less significant and the efficiency of the accelerator is determined by the value η_N which in turn depends on the phase properties and vertical focusing.

The next conclusion is more interesting. The value K enters into Eq. (6.19) in the form of a product KN which for N = 15-30, is the order of unity. Particles are primarily lost on the first orbits, thus $\bar{W}$ does not depend on N. In the final analysis the efficiency of the mode of acceleration η_N becomes appreciable, amounting to tens of percents and getting larger as the number of orbits is increased.

Let us now consider means to increase K and η_N and at the same time increase the efficiency of the microtron. In a method proposed and used in the Lebedev Physical Institute[66] a constant bias voltage was applied to the emitter. With the use of this method, emission takes place only during the positive portion of the half period of the alternating electric field in the cavity which is larger than the bias value. This reduces the number of additional electrons and correspondingly increases the capture. This method was used in the first type of acceleration. The bias voltage was applied automatically during the emission current pulse by introducing a large resistance in the cathode circuit. Along with other improvements, this method increased the current on the tenth orbit to 110 MA, thus increasing η to approximately 25%.

This increase in the efficiency of the accelerator is accompanied, however, by a substantial increase in requirements on the emitter, since the region of capture does not change but emission takes place in a weaker accelerating field. It is thus necessary to drive the emitter to higher temperatures, which, of course, reduces its lifetime. Furthermore, the presence of a constant potential increases the danger of breakdown close to the emitter. All these reasons lead us to believe that the obtained current (110 MA) cannot be regarded as a practical operating limit.[66]

Another method for increasing the efficiency of the microtron depends on the use of acceleration modes with negative initial phases in which the region of a capture markedly increases. These modes were discussed in Section 6 of Chapter III. For the first type of acceleration this method was investigated and used on the first microtron by one of the authors of this book.[14,21] The second type of acceleration was investigated by L. B. Lugansky.[22]

Making use of negative initial phase modes permitted a noticeable increase in the accelerated current in our 17-orbit microtron.[27] On the last orbit, the current increased by 1.5 times (in going from a conventional mode to one with large capture) up to 50 MA. This current value was limited only by the microwave power available; thus for fewer orbits it was possible to have substantially larger accelerated currents. For instance, for 10 orbits the current was equal to 70 MA.

It is worth noting that in negative phase modes, the emitter works under the same conditions as in other modes, but the accelerated current is increased because of the larger region of capture.

SECTION 3 - CHOICE OF THE FREQUENCY FOR THE MICROTRON

As has already been noted several times, all microtrons, as a rule, operate in the 10-cm waveband (between 9-12 cm). The choice of this band was initially made because of the availability of microwave apparatus designed for radar. However, this circumstance, although important from the practical standpoint, is in no way a matter of principle. Actually, in radar, wide use is made of bands from 25-30 cm and from 3-5 cm, not mentioning some other bands even further away from 10 cm. In fact, the choice of the 10-cm waveband was determined by more fundamental reasons which make this waveband especially convenient for microtrons and in general for all high frequency accelerators. Recall that the majority of contemporary linear accelerators also operate in this waveband

The first, though not very important, reason is due to the size of the cavity: the diameter of the circular cavity with an E_{010} oscillation mode is approximately equal to the length of the wave, and thus, for longer waves, the cavity becomes too large while for shorter waves (for instance the 3-cm waveband) it becomes too small.

The second (and far more important reason) is the fact that for each wavelength λ there is a corresponding characteristic electric field

$$E_0 = H_0 = \frac{2\pi mc^2}{e\lambda} ,$$

which for $\lambda = 10$ cm gives a value of $E_0 = 320$ kV per centimeter. This field gradient characterizes the intensity of the fields in all microwave devides of this waveband. Thus, the field gradient in the accelerating cavity in a microtron is equal to $E_0 \Omega$.

In linear accelerators, the use of traveling waves, the phase velocities of which are equal to the speed of light, is possible only when the field gradient of these waves is close to E_0. For this reason it can be said that microtrons and linear accelerators use strong fields, i.e., fields in which particles attain the energy U_0 at a distance of the order $\lambda/2\pi$.

It is true that the potential $E_0 = 320$ kV/cm which corresponds to $\lambda = 10$ cm is close to the highest fields that can be obtained in copper cavities. The gradient is determined by microscopic effects which are in turn based on atomic field potentials. For polished copper cavities under fairly good vacuum conditions and not very high temperatures, the limiting is $\sim 10^6$ V/cm.

This is a good occasion to mention that although the problem of the high field limit in microwave fields is of great practical interest, it has not been systematically investigated. One can only conclude that the so-called electrical strength of the vacuum is determined by a series of elementary processes on the surface of metals. These processes are associated with films which always are formed on metal surfaces but which have little relation to the metal. They also depend on the properties of the metal itself. It is reasonable to expect that there would be a great interest in the study of electrical stability in pulsed and, especially, microwave fields where bombardment by ions and other processes leading to mechanical changes in surfaces similar to the growth of microcrystals would be reduced to a considerable degree. One can surmise that under such conditions the highest electrical fields will be determined more by the fundamental processes related, through auto-emission, to atomic fields, than by a series of obscure conditions which form the contradictory picture we have at present.

The limiting magnetic and electric fields which can be obtained in these or other cases have, as has already been mentioned, great practical importance. In the case of constant magnetic fields, the main technical limit is the magnetic saturation of the steel or other metals used, and the microscopic mechanism that leads to magnetic field limitation in ferromagnets are well understood. In the case of superconductors the limiting value of the magnetic field is also determined by the properties of the substance. In the case of electric fields, however, for which an adequate understanding of the phenomena involved does not exist, we are faced in each case with the technical limits imposed by a certain device, be it the column of a Van de Graaff accelerator, the waveguide of a linear accelerator, or the cavity of a microtron.

Large microtrons, up to 30 MeV, operate in the short-wave part of the 10-cm waveband, with $\Omega = 2$. Limitations associated with electrical breakdown in the cavities of these accelerators have been observed. For this reason, the development of microtrons utilizing waves of shorter than 5 cm with $\Omega \approx 1$ appears to be impractical at this time, if one does not go to low $\Omega < 1$.*

On the other hand cavities operating in the longer wave of the S-band (with $\Omega = 1.7$) do not require special treatment and very high vacuum conditions. With further increases in the wavelength, the field strength decreases. This brings about an increase in the size of the magnet and consequently

---

*At the present time a set of microtrons is operating at $\lambda = 3$ cm with $\Omega = 0.5$-0.6.[98]

of the accelerator. For this reason making use of waves much longer than 10-15 cm in pulsed microtrons is not really very reasonable. Longer waves (15-20 cm) are more suitable for cw microtrons. In machines of this type the limiting factor appears to be in the heat dissipated in the cavity which is proportional to $\lambda^{-1/2}$ and consequently decreases with increase in the wavelength.

SECTION 4 - CONDITIONS OF SIMILARITY FOR A MICROTRON

Essentially, on looking at the various facets of microtron operation, we have actually made use of scaling conditions by writing all the equations in dimensionless form. A natural scale of length is the wavelength λ (more accurately $\lambda/2\pi$). Time, too, is the period of an oscillation T_0 of the accelerating field (more accurately $1/\omega = T_0/2\pi$). The electric and magnetic fields were expressed through the cyclotron field H_0 (or E_0).

Parameter Ω was introduced as the ratio of the guide magnetic field to H_0, or as the ratio of the energy gain per revolution to the rest energy. This parameter can also be interpreted in a slightly different way: namely, as the effective mass of the electron. In all the main equations of the accelerator dynamics the energy is introduced, not as Γ, but as the ratio Γ/Ω (see Refs. 14, 22). This means that if we were to increase the mass of the electron, m, Ω times, then all the equations for the microtron will have the same form as for $\Omega = 1$ and electron mass m. The only thing that is invariant to this kind of transformation is the condition of injection into which the actual rest mass has to enter.

Most interesting is the scaling while varying the wavelength λ. This is essential for the design of microtrons and choice of the operating frequency.

While investigating the effect of varying the wavelength we should consider two cases. In the first case, variation of λ can take place under full geometric similarity of the accelerator, where parameter Ω, the number of orbits N, the energy of the beam U_N, the geometry of the cavity, and the conditions of particle injection are unchanged. In the second case we will consider that while varying λ the parameters which remain unchanged are the beam energy and the size of the magnet. Naturally, the strength of the constant magnetic field and the alternating field in the cavity remain unchanged. The number of orbits and the parameter Ω in the second case are variable. In Table 6.1 the main microtron characteristics as functions of the wavelength are given. Of particular interest are the changes in two of the properties: the power loss in the walls of the cavity P_r and its surface density p_r.

In the first case (full geometric similarity) the power P_r is proportional to $\lambda^{-1/2}$, while in the second case, of constant fields, it is proportional to $\lambda^{3/2}$. On the other hand, the power of similar microwave generators and amplifiers is usually described by the condition $P_s \sim \lambda^2$ and thus, transition to longer wavelength always increases the radio-frequency power available to the accelerator.

MICROTRON

TABLE 6.1 Conditions of Similarity for the Microtron

	Geometric Similarity	Constant Diameter Magnet
Number of orbits N	λ^0	λ^{-1}
Magnetic field H	λ^{-1}	λ^0
Parameter Ω	λ^0	λ
Electric field E	λ^{-1}	λ^0
Power losses in cavity P_r	$\lambda^{-\frac{1}{2}}$	$\lambda^{3/2}$
Density of power losses in cavity p_r	$\lambda^{-5/2}$	$\lambda^{-\frac{1}{2}}$
Field tolerance $\Delta H/H$	λ^0	λ^{-2}
Tolerance on the shape of pole Δh	λ	λ^{-1}
Magnet weight	$\lambda^3 - \lambda^{5/2}$	$\sim \lambda^0$
Magnet aperture h	λ	λ
Magnet power requirements	λ^2	λ

In a cw microtron one of the most critical parameters is the specific loss p_r. In the case of complete geometric similarity $p_r \sim \lambda^{-5/2}$, in the case of constant field $p_r \sim \lambda^{-\frac{1}{2}}$, and we can see that p_r decreases in both cases. In cw machines the problem of heat transfer determines the high frequency limit.

The physical meaning of this limit is the following. For a certain field in the cavity, and consequently for a certain current density in the walls, p_r changes as the surface resistance of the cavity (in proportion to $\lambda^{-\frac{1}{2}}$). Thus, when the magnetic field H is approximately 500 Oe and the electric field gradient is approximately 150 kV/cm, in cw operation it is not possible to use waves much shorter than 15-20 cm since, even in this case, p_r = 250-350 W/cm^2.

In Section 3 we have already presented circumstances which limit the values of Ω and λ for pulsed microtrons associated with limits in the electric field in the cavity.

Let us emphasize once again that the magnetic field in a microtron is not the limiting factor. Analyzing the scaling in microtrons, we can consider that the magnets in these accelerators are far from being saturated and we can treat them as linear devices. In this respect the microtron differs from other cyclic accelerators for which the magnetic field determines the scaling as to be limited by saturation.

SECTION 5 - BEAM-CAVITY INTERACTION

In our microtron investigation thus far we have studied the motion of particles in a given field. At the same time, we have treated the electron beam as some constant load which receives power from the cavity but we did not take into account the effect of phase oscillations in the beam on the accelerating field in the cavity. However, at higher values of current and power in the beam these effects cannot be ignored.

Work by E. L. Kosarev[25] has shown that the interaction of the beam and the cavity can be investigated by the method of simultaneous solution of the equation of phase motion and the equation which describes the forced oscillations of the cavity under the action of the accelerated beam. The main assumption made in this work was that the amplitude and the frequency of the exiting wave were constant, i.e., the generator is stable and is isolated from the cavity.*

Let us determine the average phase $\Psi(n)$ of the accelerated bunches by the relationship

$$\cos \Psi(n) = \frac{1}{N} \sum_{N=1}^{N} \cos \Omega_k(n) , \qquad (6.21)$$

in which $\phi_k(n)$ is the phase of the bunch at the kth orbit entering the cavity at the nth period of the accelerating field. Small phase deviations of all the bunches from the equilibrium phase can be approximately written as

$$\Psi(n) = \phi_s + \psi(n) , \qquad (6.22)$$

where

$$\psi(n) = \frac{1}{N} \sum \phi_k(n) . \qquad (6.23)$$

The variation in the time $\phi_k(n)$ with the variation in the amplitude of the accelerating field in the cavity, $A(n)$, is described by the system of nonuniform difference equations

$$\phi_{k+1}(n+k+1) = \phi_k(n) + \frac{2\pi}{\Omega} \Gamma_{k+1}(n+k+1) ,$$

$$\Gamma_{k+1}(n+k+1) = \Gamma_k(n) + A(n) \cos \phi_k(n) . \qquad (6.24)$$

These equations differ somewhat from Eq. (2.11) since $\Gamma_k(n)$ designates the energy of the bunch entering the cavity on the nth period of the accelerating field after having completed the kth orbit, and the amplitude of the acclerating field $A(n)$ can depend on n, i.e., on time. For small deviations of $\phi_k(n)$ and $\Gamma_k(n)$ from the equilibrium values the system of of Eqs. (6.24) can be linearized and the solution written straightforwardly.

---

*Recently the results of Kosarev have been substantiated by Melekhin and Melekhin and Luganski.[97] This work has led to a deeper understanding of high current effects in the microtron, that depend on the highly nonlinear phase motion in this accelerator, especially prominent at high currents ($\sim$ 100 MA).

MICROTRON

The action of the electron bunches on the field in the cavity is completely determined by the average phase ψ(n) which depends on the amplitude of the accelerating field A(n). Assuming that A(n) for n ≤ 0 has an equilibrium value and for n > 0 has some other constant value, we can calculate how the function ψ(n) behaves. On Fig. 6.3 the results of calculations

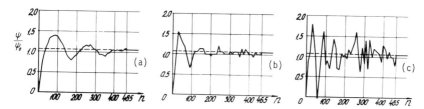

Fig. 6.3 Oscillations of the average phase of the electron beam under jump variation in the amplitude of the accelerating field: (a) $\nu = \pi/4$, $\phi_s = 5°$; (b) $\nu = \pi/2$, $\phi_s = 18°$; (c) $\nu = 7\pi/8$, $\phi_s = 31.5°$.

in the 30-orbit microtron for various values of equilibrium phase ϕ_s ($0 < \phi_s < 32°$) and the corresponding values of the frequency of phase oscillations ν determined by Eq. (2.19) are given. On the abscissa n, the number of periods of the accelerating field is shown, while the ordinate is the ratio $\psi(n)/\psi_0$, where ψ_0 is the equilibrium value of $\psi(n)$ which corresponds to the new constant value A(n) for n > 0.

Calculations (Fig. 6.3) lead us to an unexpected result: it appears that the average phase of the microtron beam Ψ(n) possesses an oscillatory degree of freedom and its small deviations from the equilibrium phase $\psi(n) = \Psi(n) - \phi_s$ approximately satisfies the second-order differential equation

$$\ddot{\psi} + 2\Delta_\phi \dot{\psi} + \Omega_\phi^2 \psi = \Omega_\phi^2 \psi_0 \qquad \text{(where } t > 0\text{),} \qquad (6.25)$$

in which the damping coefficient Δ_ϕ and the frequency Ω_ϕ depend on the value of the equilibrium ϕ_s.

Equation (6.25) is analogous to the equation of small phase oscillations in conventional phase-stable cyclic accelerators. However, the mechanism for damping the phase oscillations is different. Damping in the microtron is insured not by adiabatic damping or radiation friction, but comes from the passage of bunches through the cavity on all the orbits possessing, in the laboratory time scale, various frequencies of phase oscillations and experiencing phase displacement as a result.

The characteristic time that it takes to change the field amplitude in the cavity with a Q-factor Q and a frequency ω is equal to Q/ω. When the Q factor of the cavity Q ≥ (500-1000) then the time considerably exceeds $1/\Omega_\phi$ and $1/\Delta_\phi$ and consequently the oscillatory properties of the microtron beam described by Eq. (6.25) due to its interaction with the cavity, do not in fact exist. But in the high current microtrons when $Q_k/Q \sim 1/10$ to $1/15$ this degree of freedom is important.

The electron current loading the cavity can be divided into two components: the current caused by the accelerated bunches [it lags the accelerating voltage in phase by (n)] and the current caused by the nonresonance electrons (in the first approximation, this can be considered to be in phase with the cavity voltage). Let us designate

$$\eta^{(1)} = \frac{P_N}{P_r}, \quad \eta^{(2)} = \frac{P_e - P_N}{P_r}, \quad (6.26)$$

where P_r is the power loss in the cavity, P_N is the power in the accelerated electrons, $P_e - P_N$ is the power in the remaining electrons (see Section 2 of this Chapter). It appears (Ref. 25) that in the stationary mode, the cavity and the beam have a Q-factor

$$Q_1 = \frac{Q}{1 + \beta + \eta^{(1)} + \eta^{(2)}} \quad (6.27)$$

and a resonant frequency ω_1 determined by the equation

$$\omega_1^2 = \omega_0^2 \left(1 - q + \frac{\eta^{(1)}}{Q} \tan \phi_s \right), \quad (6.28)$$

where Q and ω_0 are the Q-factor and the internal frequency of the cavity without taking into account its interaction with the electrons or the waveguide, β is the coupling coefficient between the cavity and the waveguide, and q is the frequency correction term due to this coupling.

Equation (6.28) can be easily derived if we consider an equivalent system of a beam-loaded cavity. The beam is treated[12] as a supplementary current I_N (Fig. 6.4). Since

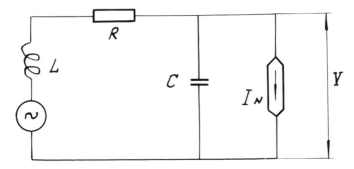

Fig. 6.4 Equivalent circuit of the beam-loaded resonance cavity.

the current phase lags the voltage V by ϕ_s, it lags the displacement current through the capacitor C by $\pi/2 + \phi_s$. Thus, the presence of the current I_N is equivalent to a change in the capacitor C by

$$\Delta C = -\frac{1}{\omega}\frac{I_N}{V}\sin\phi_s .$$

Since the power in the beam of accelerated particles is

$$P_N = \frac{1}{2} I_N V \cos\phi_s ,$$

and the power loss in the cavity

$$P_r = \frac{1}{2}\omega C/Q V^2 ,$$

we have

$$\eta^{(1)} = \frac{P_N}{P_r} = \frac{Q}{\omega C}\frac{I_N}{V}\cos\phi_s , \quad \frac{\Delta(\omega^2)}{\omega^2} = -\frac{\Delta C}{C} = \frac{\eta^{(1)}}{Q}\tan\phi_s ,$$

which leads to Eq. (6.28). Let us emphasize that this conclusion is based on the fact that the oscillating characteristic of the beam determined by Eq. (6.25) due to its interaction with a cavity is not taken into account.

As can be seen from Eq. (6.28), the resonant frequency of the cavity with the beam is higher than the internal frequency of the cavity without the beam and the increase in the frequency is proportional to the product of the corresponding power in the accelerated beam $\eta^{(1)}$ and $\tan\phi_s$, the tangent of the equilibrium phase. As the amplitude of the accelerating field is changed, the equilibrium phase of the beam also changes according to

$$A(t)\cos\phi_s(t) = \text{const} . \qquad (6.29)$$

Thus the cavity is an essentially nonlinear accelerator, the resonant frequency of which depends on the amplitude of oscillations in it. The stability of such a system has been investigated by standard methods (see Ref. 25). We will only present a simple explanation of the instability mechanism.

Let us assume the power in the beam suddenly decreases (or increases). Then the natural frequency of the cavity will decrease (or increase). For a positive detuning ($\omega - \omega_1 > 0$) when the frequency of the exciting generator ω is greater than the resonant frequency of the cavity and the beam, a change in the power of the beam causes an increase (a decrease) of the detuning. Because of this, there is a drop (a rise) in the amplitude of the accelerating field and, correspondingly, in the power of the beam. The net result of this is that starting with a relatively large initial current in the beam, the equilibrium phase goes beyond the limits of the stable region and acceleration ceases. A similar argument demonstrates that for a negative detuning ($\omega - \omega_1 < 0$) the process of acceleration is stable for any current.

According to the above considerations, the interaction of the beam with the accelerating field in the cavity of the microtron should not limit the current at least up to the point where the beam loading does not become so large that the cavity and the beam can no longer be treated as a resonant system.* The determining factor here is the ratio

$$\tilde{Q} = \frac{P_r}{P_e + P_r} \quad . \tag{6.30}$$

As long as it is large, the above argument is valid. According to Eq. (6.7) $P_r \sim 2000$ MW, and as long as the beam power is considerably less than this value the theoretical results presented can be used.

At the same time, we should point out that in the theory presented, phase oscillations are treated in the linear approximation. We should caution that for beam power where Eq. (6.30) is still large, real perturbations cannot be considered small since they could destroy the mode of acceleration.

The transient processes associated with the establishment of oscillations in the cavity and the establishment of acceleration in the accelerator depends to a considerable degree on the properties of the microwave generator and its coupling to the cavity. We will not investigate these processes.

As we have already mentioned, because of the nonperiodicity of the phase oscillations for a relatively large number of orbits the beam is an aperiodic system. A strong stabilizing factor is the noticeable dependence of the current in the beam on the accelerating voltage. This is associated with the dependence of the emitted current on the field during beam injection directly into the cavity. For a constant injected current, with an external gun, for instance, the power in the beam, as has been shown above, does not depend on the accelerating voltage. In that case, the differential conductance of the beam is equal to zero.

During the motion of electrons through the cavity, it is also possible to excite higher frequency oscillations. Taking beam bunching into account, it may be considered that the beam effectively contains the first 8-10 harmonics of the current. Excitation of the cavity by these is possible only when the harmonic frequency coincides with the resonant frequency of the cavity. Electron bunches moving along the axis of a circular cavity will effectively excite oscillations E_{0n0}, in which the electric field is not equal to zero at the axis (also oscillations E_{01p} and to a smaller degree E_{02p}, E_{03p} and so on: $n, p = 1, 2, 3$).

The spectrum of the frequencies E_{0n0} is determined by the roots ν_{0n} of the equation $J_0(\nu) = 0$. The frequencies of this spectrum are not harmonics of the fundamental frequency, and so there is no direct danger of such frequency coincidence in a circular cavity. If we have a rectangular cavity, where the electrons are accelerated by a field operating in the E_{11c} mode, then the electron bunches will excite the frequency E_{nm0} in the cavity and also the frequencies E_{nm1}, E_{nm2}, etc. The natural frequency E_{nm0} is equal to

---

* Nonlinear phase resonance can cause high current instability at very little current (~ 1 MA).[97]

MICROTRON 147

$$\omega_{nn0} = n\pi \sqrt{2}\, \frac{c}{a}\, n\omega_{110} \quad,$$

i.e., it will be the nth harmonic of the fundamental frequency.
It is possible to excite the frequency E_{110} in a circular cavity when the position of the beam is displaced from the axis. Excitation and amplification of such oscillations takes place in linear accelerators and causes loss of the transverse stability of the beam for currents greater than a certain value. It may be expected that for a microtron with a relatively large current there will also be a loss in radial and vertical stability. We must consider, however, that because of the large size of the aperture coupling the higher harmonics will be substantially loaded by the waveguide, but the real danger due to the excitation of the higher number harmonics can only be determined experimentally.

SECTION 6 - COHERENT RADIATION OF THE BUNCHES AND THE LIMITING CURRENT IN THE MICROTRON

Let us investigate coherent radiation of electrons during the motion of bunches through the microtron following the work of S. P. Kapitza and L. A. Wainstein.[6]
During circular motion in the magnetic field of an accelerator, electrons radiate but this noncoherent radiation is so small (of the order 10^{-3} eV/revolution) that it can be completely ignored. However, at wavelengths of the order of the bunch dimension, and shorter, the electron bunches radiate coherently. This increases the radiation of each electron by N times, where N is the number of electrons in the bunch. Let us evaluate this effect.
Power radiated by a charge during circular motion along a radius R in a magnetic field H, is determined by the expression

$$P = 2/3\, \frac{e^4 H^2 \beta^2 \Gamma^2}{m^2 c^3} = 2/3\, P_0\, \frac{r_0^2}{R^2}\, \beta^4 \Gamma^4 \quad . \tag{6.31}$$

We will consider the motion as being ultra-relativistic ($\beta \approx 1$). Let us designate ω_1 as the frequency of revolution for a given energy Γ, and ω as the frequency of the accelerating field. Then the time for one revolution is

$$\tau = \frac{2\pi}{\omega_1} = \frac{2\pi\Gamma}{\omega\Omega} \tag{6.32}$$

and an electron will lose energy

$$\delta U = \frac{4\pi}{3}\, \frac{e^3 H}{mc^2}\, \Gamma \quad . \tag{6.33}$$

This energy is radiated in the form of a quasi-continuum spectrum, with limiting frequency of the order of

$$\bar{\omega} = \omega_1 \Gamma^3 = \omega\Omega\Gamma^2 \quad . \tag{6.34}$$

The effective dimension r_e of the bunch is determined by the region of phase stability and according to the investigation of Bikov[7,8] (see also Chapter VIII), we must consider

$$r_e = \frac{\lambda}{2\pi S} , \qquad (6.35)$$

where $S = 8$ is the beam bunching coefficient. We can obtain the total energy loss of the electron by multiplying Eq. (6.33) by the number of particles in the bunch, determined by the equation

$$N = \frac{2\pi J}{e\omega} \qquad (6.36)$$

(J is the average current in the pulse), and by the coefficient of coherentness Θ ($\Theta < 1$).

In Ref. 6 there is a detailed calculation for an ultra-relativistic spherical bunch. It was shown that Θ is the function $(r_e/R)\Gamma^3$, where r_e is the effective dimension of such a bunch and R is the radius of the orbit [according to Eq. (2.9) $R = (mc^2/eH) \times \beta\Gamma \approx (mc^2/eH) \times \Gamma$]. We have

$$\Theta \approx 1 \quad \text{for } \frac{r_e}{R} \Gamma^3 \ll 1 , \qquad (6.37)$$

and

$$\Theta \approx \frac{\kappa}{(r_e/R)^{4/3} \Gamma^4} \quad \text{for } \frac{r_e}{R} \Gamma^3 \gg 1 , \qquad (6.38)$$

where κ is a numbering coefficient. For a uniform bunch (a sphere of radius r_e) $\kappa = 1.43$, and for a Gaussian bunch (the charged density is proportional to e^{-r^2/r_e}, where r is the distance from the center), $\kappa = 0.53$.

The change in the dimensionless energy Γ during one revolution caused by radiation is determined by the expression

$$\delta\Gamma = \frac{8\pi^2}{3} \frac{J}{I_0} \Omega\Theta\Gamma^2 , \qquad (6.39)$$

where I_0 is the characteristic electron current (6.5). The corresponding change in the arrival phase is equal to

$$\delta\phi = \frac{\pi}{\Omega} \delta\Gamma . \qquad (6.40)$$

If the energy $\delta\Gamma$ was lost at the start rather than during the complete orbit, then the factor π in Eq. (6.40) would have been replaced by the factor 2π.

From the theory of phase motion, it follows that $\delta\phi$ cannot be of arbitrary value without destroying the process of acceleration. If the phase at the start was optimal [see Eq. (2.24)], then we can consider

MICROTRON

$$\delta\phi_{max} = 0.15 \quad . \tag{6.41}$$

A more detailed argument for this is given in Ref. 6.
In this way we can determine the maximum current

$$J_{max} = \frac{3I_0 \delta\phi_{max}}{8\pi^3 \Theta \Gamma^3} \tag{6.42}$$

at which the energy loss from radiation displaces the phase by $\delta\phi_{max}$ during one revolution.

For $\Omega = 1$, $S = 8$ and $\Gamma = 3$ (on the second orbit) we can make [according to Eq. (6.37)] $\Theta = 1$ and obtain $J_{max} = 9$ A, i.e., even on the first several revolutions, coherent radiation begins to limit the current.

For large energies, the spectrum of radiation is displaced in the direction of higher frequencies, where the electrons radiate noncoherently ($\Theta < 1$) and we must make use of Eq. (6.38) in place of Eq. (6.37) or a supplementary expression that is more complicated (see Ref. 6). Equation (6.38) can be derived by taking an expression for the radiation at frequencies considerably lower than the limiting frequency (6.34) and integrating (summing) this radiation from $n = 0$ to $n = S\Gamma/\Omega$. Proceeding this way we obtain only a small change in the value of κ.

Making use of Eqs. (6.42) and (6.38), and for the same parameters as above ($\Omega = 1$, $S = 8$), and an energy of 20 MeV ($\Gamma = 40$), we obtain $J_{max} \approx 1$ A, i.e., the current is an order of magnitude smaller than that given above. At higher energies, the limiting current J_{max} slowly decreases (inversely proportional to $\Gamma^{1/3}$). We must note the increase in the limiting current (proportional to $\Omega^{4/3}$) with an increase in the magnetic field. On Fig. 6.5 the dependence of the limiting

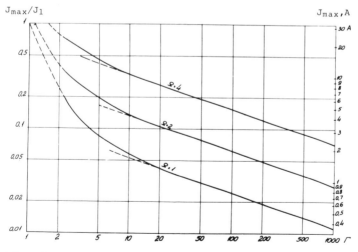

Fig. 6.5 Limit on current in the microtron ($J_1 - 32$ A).

currents on Γ and Ω for a Gaussian bunch is presented. On these curves, one can clearly see the initial region where Eq. (6.42) can be used for $\Theta = 1$ and the region of large energies, where the limiting current falls inversely proportional to $\Gamma^{1/3}$.

A complete investigation of the motion of electrons taking into account coherent radiation and its effect on the beam thus far has not been performed. The considerations presented are limited to the value of current at which these phenomena start to appear.

These evaluations are approximate because of the fact that for $J \approx J_{max}$ the shape of the bunch and the charge distribution in it begins to be affected by Coulomb and radiation forces. The effect of the negative mass instability can account for squeezing of the bunch in the longitudinal direction, which in turn can change the character of the radiation. In Ref. 6 it was shown that the coefficient of coherentness Θ for spherically symmetric (in the laboratory system of coordinates) bunches and for a linear (worm type) bunch spread out along the axis of revolution is the same if the radial dimension of the spherical bunch is equal to the length of the linear bunch. It follows from this that the radial dimension of the bunch does not affect the coefficient of the bunch. Thus, when the bunches lengthen in the longitudinal direction, radiation increases.

In calculating the radiation from the bunches, we have ignored the effect of the walls of the chamber. However, the distance to the poles of the magnet is of the same order as the wavelength of the first harmonic radiated by the bunch, and thus the effect of the wall is not large. Closely placed walls will, generally speaking, reduce radiation. Close to the cavity itself, there will be interaction between bunches on different orbits, however the role of this interaction is not clear. It should be mentioned that the net current on the orbits destroys the uniformity of the magnetic field and, at the same time, the process of acceleration. Such an interaction between the dc component of the electron current and the magnetic field can also cause limitation of the current

Equation (6.42) must be used very carefully for evaluating the current in which the collective effects can play a known role in the dynamics of the accelerator. This is evident at least from the fact that by using a drooping magnetic field it is possible, within known limits, to compensate radiation of the bunch as a whole, and at the same time to increase the limiting current (see Ref. 23).

CHAPTER VII

EXPERIMENTAL STUDY OF THE MICROTRON

SECTION 1 - POWER MEASUREMENT IN THE CAVITY AND THE CURRENT OF ACCELERATED ELECTRONS

Starting up the accelerator begins with the establishment of the magnetic field value and tuning the cavity to the frequency of the magnetron. Tuning is determined by maximizing the value of the high frequency signal extracted by a loop from the cavity or else by the shape of the waveform reflected from the cavity (by the signal from the directional coupler situated before the cavity[73]). A more simple but less convenient method of control consists of measuring the power dissipated in the cavity. For a resonant cavity this power is at a maximum.

After the cavity is tuned, the power to it is increased and the accelerator enters the operating mode. Electrons can be accelerated if the exciting power in the cavity is higher than a certain threshold value which is determined by the power due to losses in the walls of the cavity (for the field which corresponds to the selected acceleration mode). By gradually increasing the power in the cavity and loading the cavity by the emitted current (increasing the heating of the cathode) we obtain a maximum accelerated current. When this takes place the power level in the cavity and the value of emitted current must have reasonably corresponding values and only then will there exist resonance acceleration of electrons. The establishment of acceleration can also be noted by observing the intensity of the accelerated beam or by an ionization chamber that measures the radiation background.

The best mode of acceleration can be more accurately established by observing the shape of the accelerated beam pulse. If the emitted current is low but the field in the cavity is quite high, then electrons are accelerated only at the start and at the end of the pulse (Fig. 7.1a). When there is an exact match between current and field, the accelerated current pulse has a rectangular shape (Fig. 7.1b). In this case, acceleration takes place under the condition of equilibrium phase. When the current is too large, the electrons are either not accelerated at all or they are accelerated only in the central part of the pulse (Fig. 7.1c) when the electric field just reaches the value needed for acceleration with the equilibrium phase close to zero.

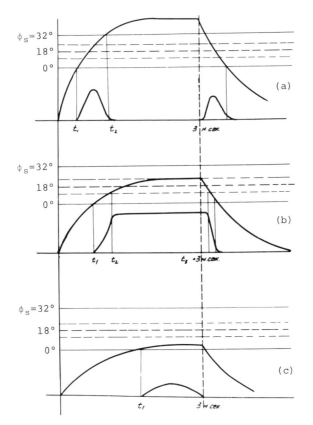

Fig. 7.1 Oscilloscope trace of the beam current as a function of the field strength, E, in the resonance cavity.

After the operating mode is established the next step is to retune the cavity, the frequency of which changes as it warms up. After a short period of time associated with the establishment of thermal equilibrium in the accelerator, the cavity operates in a stable mode with practically no attention from the operator.

Power dissipated in a cavity (it consists of ohmic losses and the power from the electrons lost in the walls of the cavity) is measured by a calorimeter as shown in Fig. 5.25 (see Section 9, Chapter V). Measurements have indicated that thermal losses due to radiation and heat transfer account for 5-10% of all the losses. Power losses in the walls of the cavity which correspond to the selected mode of acceleration can be conveniently determined by the following method.[14] The measurement is made of the power dissipated in the cavity as a function of the emitted current (for a constant field amplitude). Field stability in the cavity is controlled by

MICROTRON

observing the beam pulse on an oscilloscope: its shape must correspond to a constant and optimum equilibrium phase (Fig. 7.1b). For low values of beam current it is difficult to control the shape on the oscilloscope. For this reason, one must find the threshold power by extrapolating the obtained dependence to zero current.

On Fig. 7.2 is shown the dependence of the power on the current for the modes of acceleration which correspond to various values of the parameter Ω (which is proportional to the

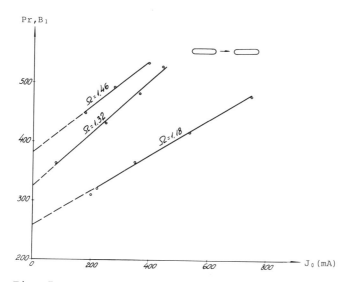

Fig. 7.2 The dependence of power dissipation in the cavity on the emitted current (cylindrical cavity $\ell = 1.2$).

magnetic field and the energy gain per revolution). In all cases the dependence appears to be linear. This means that the relative losses of electrons on each orbit are the same for various currents. Consequently, the action of space charge and other strong current effects do not appreciably influence the value of electron loss. This conclusion is verified by the fact that the accelerated current is proportional to the emitted current, i.e., the coefficient of capture is practically independent of the strength of the current, as long as the current is not too large. In practice, with proper precautions, high current effects become observable at currents $\sim$ 50-100 MA, although beam loading and high current instability effects can be observed at low current ($\sim$ 1 MA in an unfavorable condition).[97]

The indicated method permits experimental determination of the efficiency of the mode of acceleration η_N (for the determination of η_N see Section 2 of Chapter VI). In Refs. 14 and 21, the value η_N was measured for modes with zero initial phases and for an effective mode (the first type of acceleration) for which the calculated characteristics were presented in Chapter III. Measurements were performed on our 12-orbit microtron. The power of the accelerated beam, P_N, was determined by the electron current and energy. The power expended to accelerate electrons lost on the walls of the cavity was determined by subtracting the ohmic losses from the total power dissipated in the cavity. Measurements have shown that for a mode with initial phases close to zero, η_N amounts to 20-25% and for the effective mode (negative initial phases) η_N was approximately 45%.

The energy of accelerated particles can be varied within wide limits (while using the first type of acceleration) if the apertures are in the shape of long horizontal slits. By varying the constant magnetic field and the power in the cavity we move to modes of acceleration with variable energy gains per revolution as a result of which the particle energy at each orbit also changes. With this, the exit coordinate of the electrons on the first orbit also changes, which can only take place if the flight apertures are long horizontal slits.

On Fig. 7.3 the dependence of the coefficient of capture (on the 12th orbit) on the value of the magnetic field as

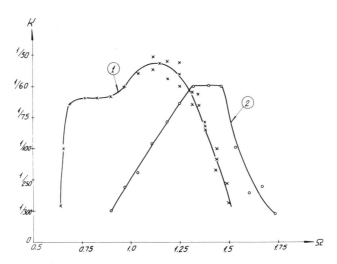

Fig. 7.3 The dependence of the capture on the parameter Ω.

measured on our first microtron is presented (see Ref. 14). Curves 1 and 2 correspond to various positions of the emitter. If the emitter is placed further away from the axis of the cavity (curve 2) then acceleration can take place in a stronger magnetic field. As can be seen from Fig. 7.3 particle energy on the 12th orbit can be continuously varied by a factor of approximately 2.3 (curve 1).

The dependence of the beam current on the number of orbits is of interest. This dependence can be most easily observed by automatically monitoring the current on a potentiometer while moving a probe from orbit to orbit (see Fig. 7.18). It can be clearly seen that after the first 2 or 3 orbits, the beam is completely formed and remains practically unchanged until the last orbit. Small variations in the current are caused by radial and vertical oscillations of the beam. It must be mentioned that such curves can only be taken at small values of beam power when the beam on the outer orbits does not appreciably load the cavity.

The position and the shape of the orbit can be determined with the aid of a probe or narrow movable targets. This method, which is widely used on our microtrons, is described in Section 3 of this Chapter. O. Wernholm[59] observed the shape of the beam (the vertical cross section) simultaneously on all the orbits with the aid of a screen made of thin tungsten wires which were traversed by the beam and were strung along the common diameter of the orbits (Fig. 7.4). Another

Fig. 7.4 A vertical cross section of the beam in the Swedish microtron (Ref. 59). Distance between wires is 1.5 mm.

method for observing the orbits in the microtron was proposed and carried out by A. P. Zhukov and V. P. Stepanchuk.[75] This method consists of placing close to the median plane, a plate covered with sodium chloride or potassium chloride. This coating was exposed under the action of electrons and gamma radiation reproducing the shape of the beam in the horizontal plane (Fig. 7.5). The position of the beam relative to the median plane can be easily determined by the current on the two halves of a target split along the median plane.[89]

SECTION 2 - THE STRUCTURE OF THE ELECTRON BUNCHES

Electron bunching was investigated by V. P. Bikov[7,8,13] who measured charge distribution in electron bunches formed in our first microtron. Charge distribution in bunches is of interest for relativistic electronics and also for improving our knowledge of phase motion.

Charge distribution in bunches was measured by microwave analyzer (Fig. 7.6) on the 12th and 11th orbits (at energies 7.3 and 6.7 MeV respectively). The scanning element consisted of a toroidal cavity (Fig. 7.7) in which the electric field was directed perpendicular to the orbits. This cavity was tuned to the frequency of the magnetron with the aid of a differential screw mechanism.

The scanning cavity was attached to a rigid coaxial line. The coupling between the line and the waveguide was determined by the length of the probe placed in the waveguide. The cen-

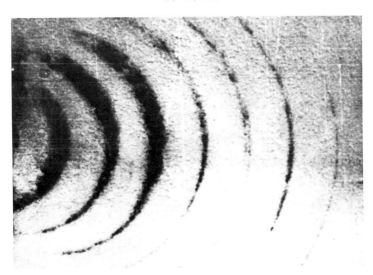

Fig. 7.5 A print of the beam obtained with the aid of a rock salt plate (the Saratov University microtron, Ref. 75).

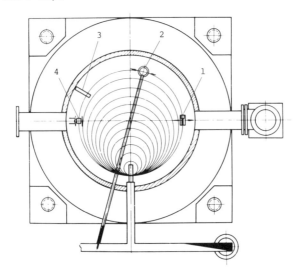

Fig. 7.6 Schematic of the bunch analyzer: (1) Entrance diaphragm; (2) Scanning cavity; (3) Screen with a horizontal slit; (4) Faraday cup.

tral conductor of the coaxial line and the scanning cavity were water-cooled.

The operation of the analyzer can be best understood from Fig. 7.8. It is obvious that the electrons which are not

MICROTRON 157

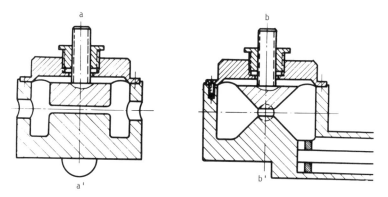

Cross Section through bb'   Cross Section through aa'

Fig. 7.7 Scanning cavity (a) section B-B'; (b) section A-A'.

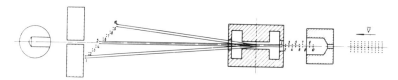

Fig. 7.8 The motion of particles in the bunch analyzer.

diverted are those which pass through the scanning cavity at such a phase that its effective field is equal to zero. By varying the phase of the field inside the cavity is is possible to change the phase of that portion of the electron current in the bunch which remained undiverted. Essentially, this arrangement is somewhat like a stroboscope in which the difference in the frequency is the rate at which the phase of oscillations in the cavity changes with respect to the phase of the beam.

In order to measure current distribution with this kind of an arrangement, it was necessary to vary the phase of oscillations in the scanning cavity relative to the phase of the bunches. To do this the scanning cavity was displaced along the orbits by bending the coaxial line. In this way it was possible to measure current distribution as a function of the position of the cavity.

The scanning cavity was displaced within determined limits by a synchronous motor. Measurement of the current passing through the diaphragm and the cavity was accomplished with the aid of a fast registering device H-110. A series of curves obtained for different positions of the entrance diaphragm makes it possible to draw contours of charge density distribution in the bunch. This method was used for investigating charge distribution on the 11th and 12th orbits of the microtron while operating in the first type of acceleration (Fig. 7.9).

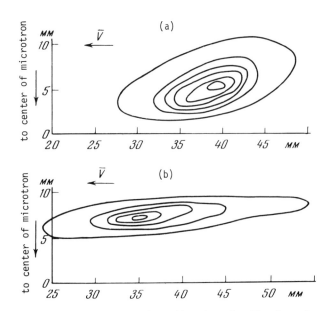

Fig. 7.9 Charge distribution in the bunches on the (a) 12th, and (b) 11th orbits.

It must be pointed out that the scanning amplitude, that is, the field in the cavity only affects the resolution of the measurement technique. In fact, the scanning value is approximately 2 cm in the vertical direction when the power dissipated in the cavity amounts to 50-60 kW per pulse. The resolution capabilities of the analyzer are evaluated as being 2×10^{-12} sec in time and about 0.5 mm in distance. We must emphasize that to obtain consistent measurements and high accuracy, very high stability of the microtron is necessary.

The current distribution is averaged over the time of the pulse. If one was to photograph the oscilloscope trace of the current passing through the measuring cavity at various moments, then one would observe the change in the shape of the bunch during the high frequency pulse. These observations, details of which have been presented in the thesis by V. P. Bikov,[13] show that the charge distribution in a bunch, in fact, corresponds to the constant field in the accelerator cavity, i.e., the field which exists during the largest part of the pulse.

In the resulting measurements it was found[8] that the effective (at half density) length and width of the bunch on the 11th orbit is equal to 8.4 and 1.5 mm respectively and on the 12th orbit, 5.5 and 2.0 mm. Electron density in a bunch can be approximated by a Gaussian distribution.

Vertical dimensions of the bunches were also measured. It turned out that on the 11th orbit the vertical size of the bunch was 2.5 mm, and on the 12th orbit it was 3.5 mm. From this the electron density in the bunch is equal to 2×10^9

cm^{-3} and, in general, the number of electrons in each bunch is approximately 10^7.

Initial measurements[7] showed a noticeable double-humped distribution of the charge in the bunch (Fig. 7.10). The

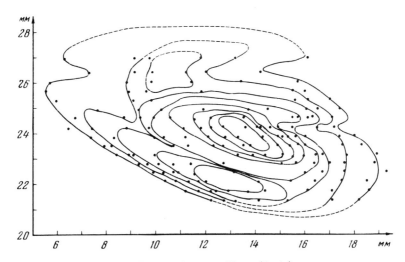

Fig. 7.10 Double hump charge distribution.

reason for such a distribution could be the two related regions of phase stability for large values of ϕ_s discovered by L. B. Lugansky.[40] A similar phenomenon was observed during investigation of radial focusing in our first microtron (see Fig. 7.31 below) and also in the 17-orbit microtron during detailed measurements performed by Yu. M. Tsipeniak.[38]*

In conclusion let us point out that measurement of bunch structure is only possible during stable operation of the accelerator. In early measurements when we used a relatively slow scope of the EPP-09 type in order to obtain complete charge distribution in the bunch the accelerator had to operate in a stable mode for periods of several hours. Subsequent measurements were performed on faster instruments and the measurement time was somewhat reduced. However, in order to obtain useful and accurate results which are in agreement with calculations, accelerator stability during operation is imperative.

Stable operation of the microtron was also necessary for the study of focusing which we will be discussing next.

---

*At present it seems more probable to explain this double-humped pattern of bunches by invoking the nonlinear resonance instability discovered by Melekhin that is especially dangerous at the phases at which Bikov operated. For a detailed discussion, see Ref. 94.

160   S. P. KAPITZA AND V. N. MELEKHIN

SECTION 3 - METHODS FOR ANALYZING THE FOCUSING

Beam focusing was investigated experimentally by V. N. Melekhin on the first microtron of the Physics Laboratory. The results of these investigations[14] are presented in this and the next section.

These investigations completely verified the conclusions obtained by theory. Among them, the most important one was the stability of vertical motion with optimally shaped cavity apertures.

The dependence of focusing on the shape of the apertures can be most conveniently studied if it is possible to change the shape of the aperture without changing other accelerator parameters. To do this we used conical inserts for the cavities (Figs. 7.11 and 7.12). Due to a slight difference in

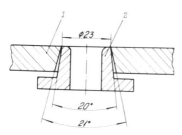

Fig. 7.11 The cover of the resonance cavity (1), with a conical insert (2).

Fig. 7.12 The target, the diaphragm and the accelerating cavity.

the cone angles between the insert and the lid, the insert fitted tightly, providing a good contact. By using many different inserts with various shaped apertures, it was possible to change the aperture without taking off the lid of the cavity and without changing the position of the emitter.

Vertical oscillations were investigated in the following manner. On the first orbit the beam was irised by the small aperture. On one of the subsequent orbits there is a copper target covered with a phosphor. The luminous spot on the phosphor caused by the accelerated electrons is observed by a closed circuit television camera. By measuring the vertical dimensions of the spot on the various orbits with a fixed position of the aperture we observe vertical motion of the part of the beam being studied, and with a different position of the aperture we investigate another part of the beam.

On Fig. 7.12 the relative position of the various parts of the structure and the arrangement of the targets is shown. Two main components of the target are the hollow copper cylinder and the brass fixture. The wall of the cylinder has a flat part, in which horizontal slits, 0.2 mm wide placed 2 mm apart are cut out. The region between the slits is covered with phosphorus. Inside the cylinder there is a small illuminating lamp which permits observation of the slits on the screen of the television. These slits made up the system used for measuring the vertical position of the electron beam.

The cylinder was attached to a brass plate which was firmly pressed to the magnet pole by the probe rod. This rod displaced the target along the common diameter of the orbits. The flat surface of the magnet pole which had a nonuniformity not exceeding 0.06 mm served as a reference plane. On account of the enlarged beam image on the television screen (by a factor of 10) it was possible to measure the vertical dimension to a high degree of accuracy (the measurements were made relative to each other and from orbit to orbit the error did not exceed ± 0.07 mm). As a result we were able to make a detailed study of vertical focusing to such a degree that we were able to observe the displacement of the neutral plane of the oscillations under the action of various perturbing factors.

The beam was irised by one of three apertures placed on one vertical line at a distance of 5 mm from each other. The upper aperture which had a 0.7 mm diameter served for locating the beam and also for controlling the operation of the accelerator by the shape of the current pulse in the target. The middle aperture, which had a diameter of 0.25 mm, served for the study of large amplitude vertical oscillations that corresponds to the peripheral part of the beam where the density of the current was small. The lower aperture, 0.1 mm in diameter, allowed us to cut out the central part of the beam and in this way permitted us to study small amplitude oscillations. The diaphragm was capable of displacement both vertically and horizontally by means of a mechanism which is shown on Fig. 7.12. The mechanism was actuated remotely with the aid of a selsyn motor.

During the study of vertical oscillations, we either used a conventional lanthanum boride emitter or a special tungsten emitter which consisted of a strip of tungsten 1.5 mm wide and 0.05 mm thick. The surface of the tungsten emitter was slightly curved. This increased the uniformity of current distribution in the beam cross section and in this way aided in the study of small amplitude vertical oscillations.

The frequency of vertical oscillations depends on the amplitude of the electron phase oscillations. For this reason, that portion of the beam which performs large phase oscillations gives a washed-out image on the target, even when a very small aperture is used. Because of this, we primarily studied vertical oscillations of electrons which performed small phase oscillations. For this case the aperture was displaced vertically through the central line of the beam until the shape of the spot on the target was the sharpest and had the smallest dimension.

During the study of vertical oscillations, it is importnat that the operation of the accelerator be very stable for rather long periods of time. The time needed for observing vertical oscillations during 12 orbits was approximately 10 minutes. If during this time the mode of operation of the accelerator changes, the oscillations will be greatly distorted. A detailed study of the total oscillation picture required several hours.

Measurement results are presented in the following section and are compared with theoretical results. Theoretical calculations were performed in the following manner: the measured coordinates of the electrons at the two last orbits were taken as the primary and the vertical motion was then calculated backwards from the last orbit to the first. In this way theory and experiment were compared at the outer orbits, where it was possible to use theoretical results.

SECTION 4 - RESULTS OF THE EXPERIMENTAL INVESTIGATION OF FOCUSING

In practice the following cavity apertures were used: a round aperture at the entrance which had a diameter of 12 mm, a round aperture at the exit which had a diameter of 8 mm (the entrance aperture had a larger diameter in order to allow the beam to go through on the first orbit), slit apertures at the entrance and the exit which had the same size (8 mm width and 20 mm length). Focusing for the first type of acceleration was investigated ($\Omega = 1.18$, $\ell = 1.2$, and $x_0 = 1.40$).

Vertical oscillations with round apertures are shown on Fig. 7.13 (the dependence of the z coordinate on the number of orbits, n, is presented). Here and hereafter, the length of the vertical line on the drawing corresponds to approximately two-thirds of the vertical dimension of the glowing spot on the target. The arbitrary vertical origin was chosen. We can see that there are oscillation nodes on the 4th and the 12th orbits and the amplitude of oscillations increases noticeably with the number n.

On Fig. 7.14 are shown oscillations with two horizontal slits; the frequency of oscillations has noticeably increased and the period of oscillations consists of 5-6 revolutions. On Figs. 7.15 and 7.16 are shown oscillations with a horizontal slit at the entrance and a round aperture at the exit, and also for a round aperture on the entrance and a vertical slit at the exit. In both cases the period of oscillations

MICROTRON 163

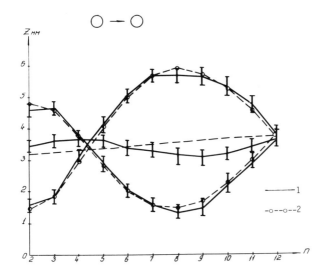

Fig. 7.13 Vertical oscillations with the use of circular apertures: (1) Experimental data; (2) Theoretical data.

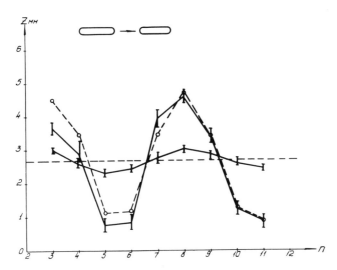

Fig. 7.14 Vertical oscillations with the use of horizontal slits. Notations are the same as in Fig. 7.13.

is approximately equal to four revolutions, which corresponds to an optimum vertical focusing. Figures 7.13 to 7.16 show that in all the cases investigated, there is a good agreement

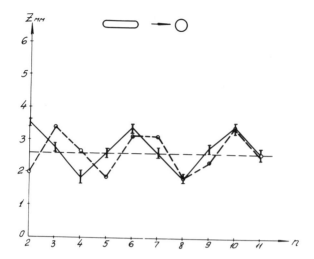

Fig. 7.15 Vertical oscillations with the use of a horizontal entrance slit and a circular exit from the accelerating cavity. Notations are the same as in Fig. 7.13.

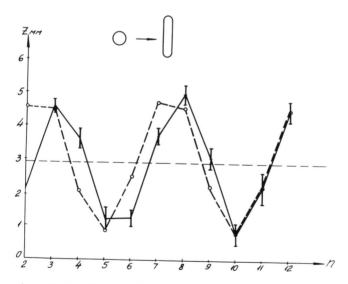

Fig. 7.16 Vertical oscillations with the use of a circular entrance and a vertical exit slit. Notations are the same as in Fig. 7.13.

between theoretical and experimental curves at the outer orbits. One can see that calculations are valid starting at about the 5th orbit.

MICROTRON

On Fig. 7.17 are given vertical oscillations for the case of a horizontal slit at the entrance and a vertical slit at

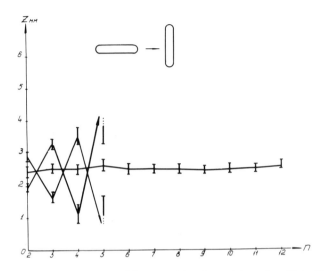

Fig. 7.17 Unstable vertical motion in the case of crossed slits.

the exit. It can be seen that the coordinate z changes sign on each revolution and grows in absolute value. This case is unstable ($S < -2$), and the middle points of the orbits lying on the common diameter are close to oscillation nodes. For this reason the vertical coordinate of the particles at these points must be small, which is exactly what we see on Figs. 7.17 and 7.30.

On Figs. 7.18 and 7.19 the current distribution on the orbits for various configurations of cavity apertures is shown. Figure 7.18 shows stable acceleration with two horizontal slits: current drop-off on the orbits is negligible. It is worth noting that Fig. 7.18 was obtained with the lanthanum boride cathode positioned in line with the internal surface of the plate; on Fig. 7.19 current distribution with the cathode recesses 0.2 mm is shown. In both cases, current distribution at the outer orbits is the same; after the 4th orbit there are no losses. Insetting the cathode produces a noticeable decrease in the losses on the first orbit and, consequently, an increase in the coefficient of capture from 1.5 to 2 times. As can be seen from Fig. 7.19, a combination of an entrance round aperture and an exit horizontal slit causes defocusing: the beam is completely lost by the 7th orbit. Crossed slits also cause instability ($S < -2$) and the beam on the first orbits is lost very soon (see also Fig. 7.17). However, on the outer orbits the drop-off becomes very slow. This fact is in complete agreement with calculations according to which vertical motion on the outer orbits

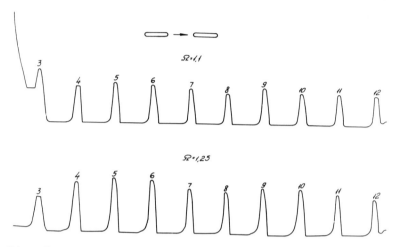

Fig. 7.18 Current distribution on the orbits with horizontal slits.

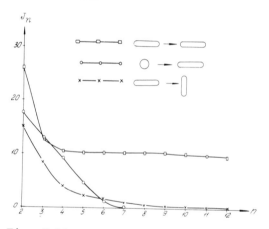

Fig. 7.19 Current distribution on the orbits with different combinations of apertures.

is stable while on the inner orbits, terms of the order $1/n$ bring about instability ($S < -2$).

We investigated vertical motion stability for various perturbations while using two round apertures and two horizontal slits. Two types of perturbations were investigated: the action of the magnetic field produced by the cathode heating current and the effect of cavity misalignment. The heating current linked the orbits and thus its magnetic field had a rather noticeable effect on the process of acceleration. By moving the phase of the current relative to the high frequency pulse it was possible to change the disturbing magnetic

MICROTRON

field by value and direction while holding the amplitude of the current constant.

On Figs. 7.20 and 7.21 is shown the perturbation of vertical motion in the case where the heater current over the

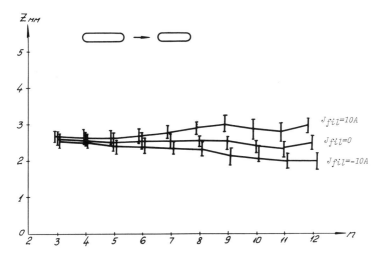

Fig. 7.20 Vertical displacement of the orbits under the action of the magnetic field created by the filament current (horizontal slits).

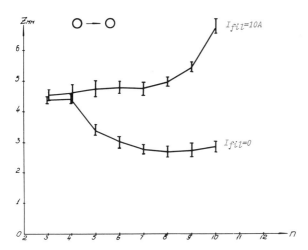

Fig. 7.21 The perturbing action of the filament current with circular apertures.

duration of the pulse was 10 A. For horizontal slits, the perturbation caused a small displacement of the beam (on the last orbit the displacement was 0.5 mm). With round apertures there is a large increase in the displacement which on the last orbit consists of 4 mm.

The effect of cavity misalignment can be seen on Figs. 7.22 and 7.23. The cavity was purposely inclined approximately

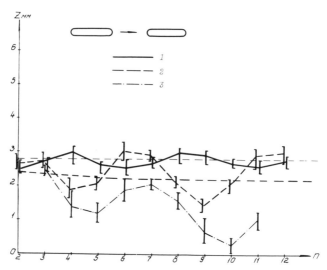

Fig. 7.22 Displacement of the orbits due to cavity misalignment and filament current (horizontal slits): (1) Undisturbed motion; (2) Misaligned cavity; (3) Misaligned cavity and filament current effects (30a).

one-half degree, i.e., its axis was inclined to the median plane as much as the clamping bolts would allow. With horizontal slits, this caused a small displacement of the oscillations (Fig. 7.22), but with round apertures the beam quickly disappeared (Fig. 7.23). In the first case, cavity misalignment causes some drop in the current on the orbits (Fig. 7.24); however, electrons are accelerated to the last orbit at which the current decreased by approximately a factor of two because of the perturbation. In the second case the beam is quickly lost and never reaches the last orbits (Fig. 7.25).

On Figs. 7.22 and 7.23 vertical motion is shown which had been perturbed by the combined action of cavity misalignment and a maximum heating current (30 A). We can see that with horizontal slits the particles are accelerated to the very end, while with circular apertures they are lost on the fourth orbit.

In this way, observations have shown, in agreement with theoretical predictions, that vertical motion with respect to various perturbations is stable for the properly selected shape of cavity apertures, while for round apertures vertical motion is strongly affected by the action of the perturbation.

Let us present the results of measurements of vertical current distribution in the beam at the nodes and anti-nodes of the vertical oscillations. We used lanthanum boride

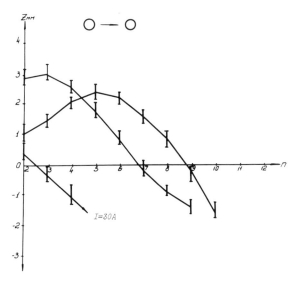

Fig. 7.23 The effect of cavity misalignment and filament current (circular apertures).

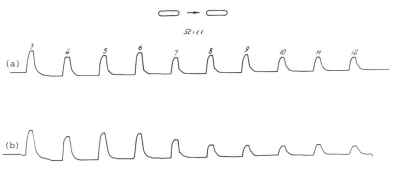

Fig. 7.24 The effect of cavity misalignment on the current decline with increasing orbit (horizontal slits): (a) Aligned cavity; (b) Misaligned cavity.

cathodes 1.5 × 1.5 mm in size, recessed 0.2 mm. Measurements were performed with the aid of a diaphragm which had a horizontal slit 0.25 mm wide in it. Vertical displacement of the diaphragm was accomplished remotely and was synchronous with the displacement of a chart recorder which recorded the distribution of the current. The resulting curves are presented in Figs. 7.26 to 7.30. The numbers given on the figures and in the text refer to the width of the recorded values while the actual dimension of the beam is smaller by the width of the diaphragm, i.e., by 0.25 mm.

On Fig. 7.26 the current distribution at the oscillation nodes on the 4th and 12th orbits, and anti-nodes on the 8th

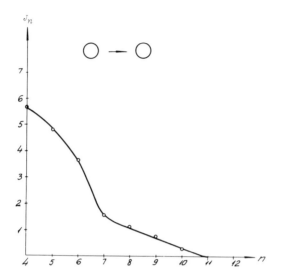

Fig. 7.25 The effect of cavity misalignment on the current decline with increasing orbit (circular apertures).

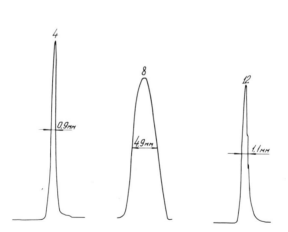

Fig. 7.26 Vertical current distribution on the 4th, 8th, and 12th orbits (circular apertures).

orbit (for two circular apertures) is shown. The width of the beam at the half-intensity level is about 1 mm at the nodes and about 5 mm at the anti-node of the oscillations. The full width of the beam on the 8th orbit is equal to 8 mm which corresponds to the vertical size of the exit aperture.

MICROTRON   171

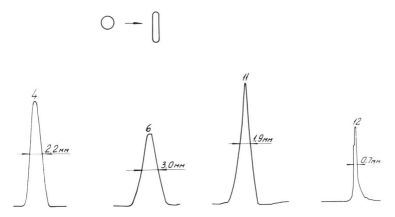

Fig. 7.27 Vertical current distribution on the 4th, 6th, 11th, and 12th orbits (circular entrance aperture and a vertical exit slit).

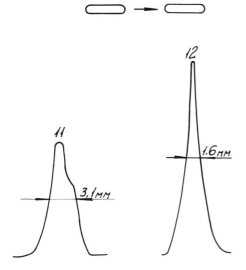

Fig. 7.28 Vertical current distribution on the 11th and 12th orbits (horizontal slits).

Let us note an interesting property of the beam in the case investigated. The small vertical size of the beam on the 12th orbit, whereas on the 8th orbit it was considerably larger, is evidence concerning the isochronism of the vertical oscillations (since the node of the oscillation washes itself out slowly). This property of the oscillations is confirmed by observation of the shape of the beam cut out by the horizontal slit.

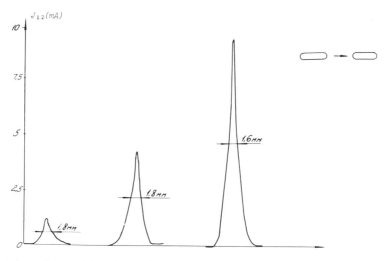

Fig. 7.29 The dependence of vertical current distribution on the 12th orbit on the current level (horizontal slits).

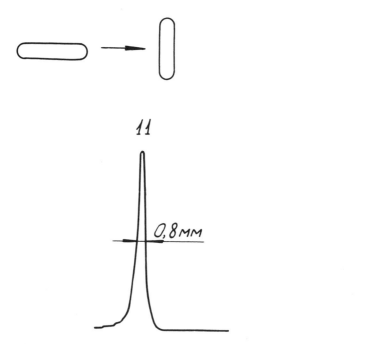

Fig. 7.30 Vertical current distribution on the 11th orbit (crossed slits).

MICROTRON

On Fig. 7.27 the vertical current distribution for a round aperture at the entrance and a vertical slit at the exit is shown. It can be seen that the vertical dimensions of the beam are small since the beam is not defocused at its first exit from the cavity (see Section 6 of Chapter III). At the oscillation node on the 12th orbit the width of the curve is about 0.7 mm, i.e., the width of the beam at half intensity is about 0.5 mm. Hence the washing-out of the oscillation node, in this case also, is insignificant. The maximum width of the beam at the anti-node on the 6th orbit is 6 mm.

On Fig. 7.28 the vertical current distribution on the 11th and 12th orbits for two horizontal slits is given. The width of the beam is 3 mm on the 11th orbit and 1.5 mm on the 12th orbit. It can be concluded that the oscillation node exists at a point somewhere between the 11th and 12th orbits.

On Fig. 7.29 experimental curves of current distribution on the 12th orbit with various beam current for the case of two horizontal slits are presented. One can see that up to a current of 10 mA there is no observed change in the shape of the curves.

The width of the beam for the case of crossed slits is small (Fig. 7.30), since the observation point, as was mentioned above, is close to the oscillation node.

From this we can see that the vertical beam dimension in a microtron is small, of the order of 3-5 mm at the antinodes of the oscillations and 0.5-1.0 mm at the nodes. Vertical angular distribution of the beam in these experiments was not measured. However, good agreement between calculated and experimental results for the z coordinate allows us to use the theoretical results obtained in Section 4 of Chapter IV. By substituting the value z_{max} which corresponds to the value of z on the 6th orbit into Eq. (4.39) (see Figs. 7.16 and 7.27), we find that the angular spread at the half intensity level for a circular aperture and a vertical slit is about 3×10^{-3} on the 12th orbit (i.e., at the oscillation node where angular spread is largest). For horizontal slits the amplitude of the oscillations is somewhat larger. However, the frequency is smaller and thus the beam has about the same angular spread. For two circular apertures the angular spread can be evaluated by superimposing Figs. 7.13 and 7.26. It reaches a maximum of 1.3×10^{-3} on the 12th orbit.

Along with the investigation of vertical focusing in Ref. 14 we also investigated certain characteristics of radial motion for various shapes of cavity apertures. On Fig. 7.31 is given the current distribution along the common orbit diameter on the 11th and 12th orbits for two horizontal slits and for two circular apertures. On the 12th orbit the width of the beam is smaller. This is due to the smaller angular spread of the beam resulting from the smaller energy spread (since the orbit is closer to the node of the phase oscillations). Taking into account the thickness of the diaphragm (0.6 mm) it can be said that the width of the beam on the 12th orbit is about 2 mm for horizontal slits and about 3.8 mm for circular apertures. It is interesting to point out that the curve of the current distribution is double-humped as we have

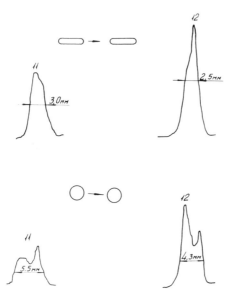

Fig. 7.31 Radial current distribution along the common orbit diameter (11th and 12th orbits).

already mentioned in Section 2 of this Chapter. By combining a circular aperture with a vertical slit we obtain a beam which is 5-6 mm wide on the 12th orbit.

Measuring the radial spread of the beam along the orbit diameter perpendicular to the common diameter of the orbits gave the following results. The width of the beam is smallest for circular apertures amounting to about 3 mm on the 12th orbit. For horizontal slits the width of the beam is 7 mm and for a combination of a circular aperture at the entrance and a vertical slit at the exit the beam is 8.5 mm wide.

It should be pointed out that for the indicated direction the orbit position is slightly nonsymmetric relative to the center of the chamber: the centers of the orbits are displaced in the direction of the cathode plate. Displacement of the center of the 12th orbit relative to the center of the chamber is about 1.8 cm for horizontal slits, 2.3 cm for circular apertures, and 3.3 cm for a combination of a circular aperture and a vertical slit.

We can determine the radial angular spread of the beam $\Delta\theta$ at points on the orbit placed along the common diameter (i.e., where the beam is extracted from the microtron), knowing the radial apread of the beam in the transverse direction. The above-given radial dimensions allow us to calculate $\Delta\theta$ by the equation $\Delta\theta = \Delta y/R$, where Δy is the radial spread, and R is the radius of the orbit. On the 12th orbit for circular apertures we get $\Delta\theta = 1.2 \times 10^{-2}$ and for horizontal slits $\Delta\theta = 2.9 \times 10^{-2}$, and for a circular aperture and a vertical $\Delta\theta = 3.5 \times 10^{-2}$.

MICROTRON   175

   These results agree with focusing calculations. Circular apertures which give weak vertical focusing provide a factor of three times less angular radial spread than a combination of a circular aperture and a vertical slit which is the best combination for vertical focusing. We can see that the use of two horizontal slits is indeed a compromise between the combinations of apertures indicated above, both from the vertical as well as the radial focusing points of view.

SECTION 5 - THE USE OF ACCELERATED ELECTRONS FOR OBTAINING
            GAMMA RADIATION AND NEUTRONS

   Many practical applications of the microtron are associated with the use of bremsstrahlung radiation. In this case the size of the electron spot on the bremsstrahlung target should be as small as possible. This is of great importance where the microtron is used for gamma radiography.
   In the last section it was shown that the vertical dimension of the beam is determined by the phase of vertical oscillations and in the final analysis depends on the shape of the cavity apertures. The latter must be chosen so that the oscillation node occurs at the target. For a large angular spread of the electrons emitted from the cavity the beam has the smallest horizontal dimension at a point on the orbit diametrically opposite to the cavity. Besides this, the orbits at this point (along the common diameter) are widely separated. For this reason, this point on the orbit is the best place to locate the bremsstrahlung target. For example, in our 17-orbit microtron the size of this spot on the last orbit at the indicated point is 2 mm high and 3 mm along the radius.
   The gamma radiation obtained is proportional to the beam current, and in a complex way depends on the thickness and material of the target, and also the energy of the beam. Thus, at an energy of 10 MeV and an average current of 50 μA in our 17-orbit microtron, the measured gamma-radiation intensity from a tungsten target 1 mm thick was approximately 2000 R/min, at a distance of 1 m from the target.[27]
   This microtron was used by Yu. M. Tsypeniuk to make measurements of gamma radiation, its spectrum and angular spread. The dependence of the intensity on the thickness of the tungsten target between 4.5 and 10 MeV is shown on Fig. 7.32. A definite maximum is observed at a thickness of about 1 mm which corresponds to 0.3 radiation lengths.
   The angular spread of the gamma radiation depends on several factors. The first is the relativistic (kinematic) spread. The approximate angle of spread is equal to $1/\Gamma$, where Γ is the relativistic factor of the electron beam. However, the kinematic angular spread can only be observed with very thin targets, of the order of less than 0.01 radiation lengths. With thicker targets a larger contribution to the spread is due to multiple scattering of electrons. Thus, at an energy of 10 MeV and an optimum thickness of the target, the angular spread of the gamma radiation is approximately two times larger than the kinematic spread. Let us also note that because of

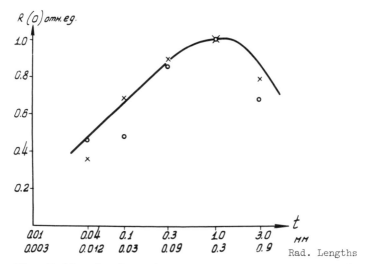

Fig. 7.32 The dependence of the bremsstrahlung intensity on the thickness of the tungsten target: 0 - W = 10 MeV; x - W = 4.5 MeV (Ref. 38).

scattering in the target, the radiation spectrum for small angles is more energetic than for large angles.

In conclusion, let us mention that angular spread of the gamma radiation also depends on the spread of the electron beam. Thus, because of the larger spread in the plane of the orbit, the field of gamma radiation from the internal target is somewhat more spread out in the horizontal direction.

The power in the bremsstrahlung radiation can be evaluated from the curve shown in Fig. 7.33. In the energy interva 5-50 MeV and a target thickness of approximately 0.3 radiation lengths, the power r at a distance of 1 m from the target is approximately equal to

$$r\left[\frac{R}{min}\right] = \sim 0.04 \; W^3 \; (MeV) \; I \; (\mu A) \quad .$$

Table 7.1 gives the properties of certain elements which can serve as bremsstrahlung targets or targets for obtaining neutrons. In the last column the threshold of the (γ,n) reaction from which neutrons are formed is given. When gamma radiation is used the formation of neutrons is usually undesirable since they activate the radiated material and the accelerator itself. One has to take this into account when using accelerated electrons at an energy higher than 10-12 MeV. Actually, when higher energies are used the cooling system becomes more complicated, necessitating the use of closed systems in order not to throw out irradiated water.

It is also worth mentioning that although the total dose of gamma radiation rises at the higher energy, the angular spread of the gamma beam decreases and consequently the radiated field is smaller. Taking all those factors into account

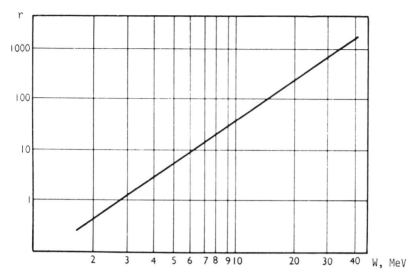

Fig. 7.33 Bremsstrahlung intensity from a thick target in roentgen units per minute at a distance of 1 mm from the target (for a 1 μA electric current).

TABLE 7.1 Properties of Elements Used for Bremsstrahlung Target.

Element	Z	Density (g/cm^3)	Radiation Mean Free Path (cm)	Thermal Cond. (J/cm sec°C)	Melting Temp. (°K)	Reaction Threshold (γ,n) MeV
Beryllium	4	1.82	28	2	1280	1.7
Copper	29	8.96	1.4	3.8	1038	10.6
Tungsten	74	19.3	0.33	1.8	3380	8.0
Platinum	78	21.4	0.29	0.7	1769	8.0
Gold	79	19.2	0.32	3.1	1063	7.9
Uranium	92	18.7	0.31	0.25	1133	6.0

we can conclude that for many purposes the electron energy of 10-12 MeV appears to be the best.

The microtron can also be used as a source of neutrons. We will not discuss this possibility in detail but for the sake of reference we will indicate the possibilities which exist here. Neutrons can be obtained by the reaction (γ,n) on heavy elements or beryllium (see Table 7.1). The neutron flux from a uranium target as a function of the electron beam energy is shown in Fig. 7.34. The uranium target here has a dual purpose, as a bremsstrahlung target and also as a converter of the gamma rays into neutrons. For purposes of evaluation we can say that for an electron energy of 25 MeV, the

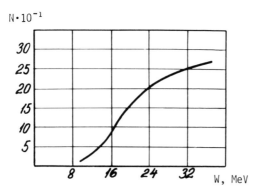

Fig. 7.34 Neutron exit from the uranium target (for a 1 kW electron beam power).

number of neutrons is 100 times less than the number of electrons, while at 30 MeV the number of neutrons is doubled. On further increases in energy, the neutron flux increases slowly and thus, 30 MeV may be considered as an optimum energy for obtaining neutrons from uranium targets with the aid of a microtron.

Recall that the accelerated electrons radiating the uranium target produced the initial burst of neutrons in the Dubna pulsed reactor of fast neutrons. Measurements on this reactor have shown that at 30 MeV the neutron flux was half that indicated above. This discrepancy may be due to the fact that the dimensions of the uranium target were not optimized.[89]

At lower energies, a uranium target is not economical; it is more convenient to use a combined uranium-beryllium target. In this target, the uranium acts as the source for the gamma radiation which is absorbed in the beryllium with the formation of neutrons. The neutron flux as a function of the electron energy is shown on Fig. 7.35.

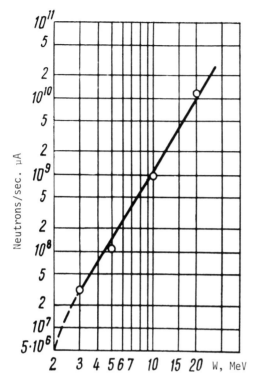

Fig. 7.35 Neutron exit from a uranium-beryllium target for small electron beam energy (Ref. 79) (for a 1 µA electric current).

CHAPTER VIII

CONTINUOUS WAVE MICROTRON

SECTION 1 - ON TRANSITION TO A CONTINUOUS WAVE MODE

Thus far, we have investigated microtrons of the pulsed variety. This is associated with the operation of those high frequency generators which excite the cavity. However, in principle the microtron as an accelerator has nothing that requires it to operate in the pulsed mode. At the present time, however, high frequency electronics has reached the kind of power level which makes it reasonable to start thinking about a continuous wave (cw) microtron. Such an accelerator with an energy of several tens of MeV and a beam power of several tens of kilowatts could be, and rightfully so, called an electron cyclotron.

A cw microtron is a very promising accelerator for nuclear physics. Classical sources of electrons for nuclear investigations such as betatrons and synchrotrons, have an intensity which is tens of thousands of times less than that which could be expected from a cw microtron. On the other hand, linear accelerators operate in the pulsed mode which is particularly inconvenient for investigations using coincidence methods. In nuclear physics, the cw microtron will undoubtedly open up new possibilities for accurate measurements and investigations of small cross sections.

In this Chapter, we will investigate the properties of cw microtrons taking into account existing high frequency generators. Powerful cw magnetrons, the so-called nigotrons,[30] proposed by P. L. Kapitza and developed in the Physics Laboratory, generate 170 kW, at a wavelength $\lambda = 17$ cm, with an efficiency of about 50%. On the other hand, within the last several years, there have been developed powerful amplifying tubes such as klystrons with a continuous power of more than 100 kW at a wavelength of $\lambda = 15$ cm with an efficiency of close to 35%, as well as powerful amplifiers of the magnetron type which possess the outstanding feature of a very high efficiency (up to 70%).

Undoubtedly, within the next several years, there will be further developments in high power electronics which will insure the development of the technology of powerful accelerators. For this reason, the problem of building a cw microtron becomes real. Initially, this problem was investigated by one of the authors.[17,18] Recently there has been another report[85] on the development of such an accelerator in Boston (USA).

MICROTRON

For various reasons of a physical nature (see Section 3, Chapter VI), it would be more suitable to operate such an accelerator in the waveband between 15-20 cm. Below we investigate the characteristics of a cw microtron at a wavelength of 20 cm. By using conditions of similarity (see Section 4 of Chapter VII) it would be easy to calculate the same properties for a different wavelength.

SECTION 2 - ENERGY CHARACTERISTICS OF A CONTINUOUS WAVE MICROTRON

Let us investigate the parameters of a microtron in which the power, P, in the cavity is equal to 150 kW and the wavelength, λ, is equal to 20 cm. At this wavelength the cyclotron magnetic field H_0 = 535 Oe. In Table 8.1 the accelerator

TABLE 8.1 Energy of the Beam and the Orbit Diameter

			Beam Energy (MeV)		
D (m)	N	D_N (mm)	$\Omega = 0.7$	$\Omega = 1.0$	$\Omega = 1.2$
2	30	1900	11	15	18
4	60	3800	21	31	37
6	90	5700	32	46	55

parameters with orbit numbers N = 30, 60, 90, and the diameter of the poles D = 2, 4 and 6 meters respectively are given. The parameter Ω has been assigned the values of 0.7, 1, 1.2. The orbit step is equal to 69 mm. Power losses, P_r, in the walls of the cavity are shown in Table 8.2, where the values

TABLE 8.2 Operating Mode of a Continuous Wave Microtron

Ω	H (Oe)	P_r (kW)	E (kV/cm)	δU (keV)	p_r (W/cm^2)
0.7	380	65	90	20	110
1.0	535	90	130	30	150
1.2	640	130	150	36	250

of the field strength in the cavity, surface density losses in the walls, p_r (which characterizes heat transfer efficiency during the cooling of the cavity), and the energy spread in the beam, δU, are also given. The value P_r is calculated according to Eq. (6.9) by using a skin depth $d = 1.75 \times 10^{-4}$ cm. For $\varepsilon = 0.8$, the dimensionless thickness of the cavity $\ell = 1.45$ and the dimensions of the cylindrical cavity were as follows: the diameter 153 mm, thickness 46 mm, internal surface S = 600 cm^2, the unloaded Q factor Q = 20,000 (for this $p_r = P_r/S$).

The value p_r is determined by the density of the heat flux dissipated by the water in the cavity. It does not exceed 250 W/cm^2 which is entirely feasible with good conditions of heat transfer. Let us note that the above-indicated high frequency generators operate with large values of p_r.

For $\Omega = 1.2$, when the power P_r is close to the total calculated power $P = 150$ kW, the energy of accelerated particles will be nearly maximum (for a given generator), and the beam current and accelerator efficiency will not be very large.

In Table 8.3 the parameters of the accelerators for the above-investigated modes of operation are given. For calculating the current in the beam and the total efficiency it was assumed that one-half of the power going into accelerating the electrons was absorbed by the beam; in other words, the efficiency of the mode of acceleration η_N (see Section 2 of Chapter VI and Section 1 of Chapter VII) was equal to 50%. This value depends on the mode of acceleration as well as the number of orbits N. As N gets larger, this value increases. However, for evaluation a more accurate value of η_N does not seem to be necessary. Let us emphasize that the indicated values η_N and other characteristics are, if anything, on the low side. The intensity of breaking radiation and the neutron flux presented in Table 8.3 were calculated on the basis of Figs. 7.33 and 7.34.

Let us point out that the critical power, which when attained, makes the practical realization of a cw microtron reasonable, is approximately 100 kW.

In figuring the power P_r we fixed the dimensions of the cavity by assuming that the microtron will operate in the first type of acceleration. It should be mentioned that it is also possible to have an acceleration mode with a smaller value of Ω for which the first revolution around the cavity will be performed during three periods of the field oscillations. In these modes m = 3 and $\Omega_{min} = 0.33$ [see Eq. (1.10) and Ref. 96]. In this way it is possible to further reduce the power (since the power P_r is proportional to Ω^2), by using cavities with still bigger values of m. It is possible that more complicated cavities could allow further reduction in the power P_r.

There is another more radical method for lowering the losses in the cavity, namely, cooling. However, at liquid nitrogen temperature (T = 77°K), the conductivity of pure copper is only 6-7 times greater and the losses in the cavity are only 2½ times less than at room temperature. At liquid hydrogen temperature there is an anomalous skin effect, after which there are no substantial reductions in the losses (the losses can be lowered no more than 5-10 times in comparison with those presented in Table 8.2).

The situation changes substantially at liquid helium temperature. Here, it would be possible to use superconducting cavities in which the losses are extremely small. For example, let us compare the losses for a cylindrical cavity with a lead coating for $\lambda = 10$ cm, $\Omega = 1.2$ and $\ell = 1.04$. The alternating electric field is equal to 400 kV/cm and the magnetic field in the cavity is 700 Oe. At room temperature the power losses in the walls of the cavity are equal to 250 kW, while

TABLE 8.3 Energy Characteristics of a Continuous Wave Microtron

D (m)	N	$P_N = 43$ kW; $\Omega = 0.7$; $\eta = 30\%$				$P_N = 30$ kW; $\Omega = 1$; $\eta = 20\%$				$P_N = 10$ kW; $\Omega = 1.2$; $\eta = 7\%$			
		U_N (MeV)	J_N (mA)	Mega-Roentgen min at 1 m	$n \times 10^{-13}$ 1/sec	U_N (MeV)	J_N (mA)	Mega-Roentgen min at 1 m	$n \times 10^{-13}$ 1/sec	U_N (MeV)	J_N (mA)	Mega-Roentgen min at 1 m	$n \times 10^{-13}$ 1/sec
2	30	11	4	0.2	1	15	2.0	0.2	2	18	0.5	0.6	1
4	60	21	2	0.7	7	31	1.0	1.0	7	37	0.25	0.3	2
6	90	32	1.3	1.5	10	46	0.6	1.5	9	55	0.17	1.0	25

at liquid helium temperature they are reduced to 3 W! The power needed to operate this refrigerator is equal to 5 kW. Regardless of the considerable amount of power in the refrigerator the gain entirely justifies it. By using longer wavelengths the losses in the cavity can be reduced further and at a wavelength of $\lambda = 20$ cm they are equal to 1 W. For this reason it would be very interesting to build a superconducting microtron. Since the constant magnetic field will not penetrate into the cavity it would be necessary to investigate the split microtron idea in which the cavity is located between two halves of the magnet. In fact, such a machine is the same type as a linotron proposed by A. A. Kolomensky.[92]

In a superconducting microtron the irising of the beam by the cavity must be eliminated because the local heating produced by the beam striking the cavity will destroy the superconductivity and force the cavity to go normal. For this reason, the problems of injecting particles into the cavity of a superconducting microtron appear to be somewhat complicated. Another problem is the stability of operation of a superconducting cavity in the absence of a beam as well as the excitation and stabilization of the field in it. These technical problems can present some serious difficulties especially in the more interesting case such as when the power in the beam is quite considerable as compared to the power losses in the cavity.

We can now conclude that a superconducting microtron is feasible. However, in discussing this problem one has to compare the superconducting microtron with a superconducting linear accelerator. Both machines will be of the continuous duty type with superconducting accelerator elements and powerful cryogenic devices. However, injecting, focusing and displacing the beam in the microtron is more complicated than in a linac. However, in a microtron it is only necessary to cool a single element, namely the accelerating cavity, whereas in a linac it would be necessary to cool all the elements along the whole length of the irised waveguide. Aside from this, the problem of cooling itself, in a superconducting accelerator, presents some serious technological difficulties.

A superconducting microtron at an energy of 200 MeV was proposed in 1965[83] by a group from Stanford University (USA). The proposal consisted of using a section of a linear accelerator of 20 MeV placed between two magnet sectors. Magnetic focusing elements would be located between the two magnet sectors. Similar proposals were published by other groups.[87] Without going into the details of these proposals, we can conclude that a superconducting linac will undoubtedly be the next step in the progress of accelerator technology.

SECTION 3 - CONSTRUCTION OF A CONTINUOUS WAVE MICROTRON AND THE POSSIBILITIES THEREOF

Let us return to the cw microtron of the conventional type (nonsuperconducting). The magnetic field in such a microtron is smaller than in pulsed machines and can be achieved with the same kind of magnets: the main parameters of such magnets are given in Table 8.4 (Fig. 8.1).

MICROTRON

TABLE 8.4 Magnet Parameters for the Continuous Wave Microtron

Pole Diameter (m)	Magnet Diameter (mm)	Yoke Thickness (mm)	Power (kW)	Weight (T)			Pole Tolerance (mm)	Pole Indicator (mm)	Deflection Due to Atmospheric Load
				Total	Pole	Yoke Cu.			
2	2300	50	4	25	7	11 0.2	0.5	1	3
4	4400	100	8	45	12	20 0.4	0.1	2	6
9	6500	150	12	110	39	36 0.6	0.05	3	9

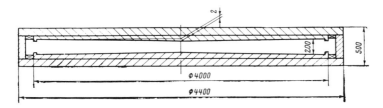

Fig. 8.1 Schematic of the magnet for the continuous wave microtron.

The distance between the poles in these magnets is taken to be equal to 20 cm. The thickness of the poles, as given in Table 8.4, was determined by assuming that the maximum induction of the yoke is equal to 13 kG. The dimensions of the magnet and primarily the thickness of the poles will be determined not by the magnetic resistance of the yoke, but by the strength of the magnet.

A cw microtron at some reasonable energy will be a machine with a large number of orbits. The larger the number of orbits the higher is the energy and the efficiency of the machine but the tolerances on the magnetic field become tighter. These tolerances are given in Table 8.4 [they were calculated according to Eq. (5.2)]. A more detailed analysis of the tolerances can be made on the basis developed in Chapter IV, namely, the theory of orbit drift and horizontal focusing. These are two ways of producing a magnetic field with the needed accuracy. The first is the accurate preparation of the poles, a method which has justified itself in small machines (N < 20). The second method is the use of correcting coils such as the ones discussed in Section 2 of Chapter V (see Fig. 5.9). It is most likely that in a large cw microtron the use of such correcting coils would be imperative. The presence of such coils will allow regulation of the position of the orbit. Correction of the orbit is especially important in powerful machines at high energies where beam losses cause not only a lower efficiency but also produce radioactivity.

In a cw microtron the magnetomotive force of the mail coils is about 10,000 ampere-turns. The system of 20 correcting coils will require a current of 1 A in each coil in order to compensate for nonuniformity which may be expected in a magnet prepared with the accuracy common to electrophysical machinery. On the other hand, it is possible to increase horizontal focusing by creating a field which is somewhat drooping from the axis of the orbits (see Section 5, Chapter IV).

From all that was said we can conclude that the necessary magnet is technically feasible.

The emitter in a cw microtron will operate in much more relaxed conditions than in the pulsed microtron because of the low current required (about 50 mA). Besides this, the larger size and the weaker field acting on the cathode will allow the selection of accelerating modes which will give large coefficients of capture and will provide more control over the process of injection. On the other hand, one should keep in mind that electron losses in a continuous wave machine will be especially dangerous since they will not only lead to an energy loss but also to local overheating of the cavity. For this reason, the processes of phasing, forming and collimating the beam will require special additional elements and accurate control methods.

In cw microtrons operating at the above-indicated currents and energies one need not expect any problems associated with space charge or radiation which could destroy the accelerating process.

We will not discuss the problems associated with the waveguide since they depend on the properties of the generator.

In conclusion we should bring your attention to the possibility of operating the microtron at a low duty ratio (about 10). In such a mode the energy rises by a factor of three for the same average machine power and the same cooling intensity. Operating a microtron in such a mode will greatly increase its possibilities and would make it feasible to have an accelerator with an energy of 100 MeV. The realization of such a mode for long pulses to a large extent depends on the presence of correspondingly suitable high frequency generators.

At the present time it would be reasonable to build a cw microtron which had a magnet 4 m in diameter and 60 orbits. At a power of 150 kW the maximum energy would be 30-40 MeV. By operating the machine in a long pulsed mode (duty ratio of 10), the same average power will raise the energy to 100 MeV.

A cw microtron is a logical extension of this type of accelerator. The main principles of its operation are the same as those of a pulsed microtron. The problem of building such a machine would only have limitations of a technical nature.

CONCLUSION

As can be seen from this book, the microtron, among other accelerators, occupies a special place since in it the particles at each revolution gain an energy which is comparable to the total energy. This special quality of the microtron means that particle injection into it takes place in a unique way (Chapter III), extraction is easy (Chapter V), and the calculations of particle motion have to be made by means of numerical methods (Chapters II and III). Particle focusing in the microtron, unlike other accelerators, is accomplished by the high frequency field of the cavity (Chapter IV). The mechanism of autophasing acting in the microtron leads to the formation of compact monoenergetic electron bunches (Chapters I, II, and VII).

The maximum energy which can be attained in a microtron depends on the allowable number of orbits and on the parameter Ω introduced in Chapter I which determines the energy gain per revolution and is equal to the ratio of the constant magnetic field to the cyclotron magnetic field which corresponds to the frequency of the accelerating field. The number of orbits is determined by the homogeneity of the magnetic field. If the homogeneity of the field is accurate to 0.01%, then the number of orbits can go up to 50 (see Chapters IV and V) while a special system of correcting coils will allow an even higher number of orbits.

For a given number of orbits and frequency it is possible to increase the particle energy by using modes of acceleration which have large values of Ω. We were able to use modes with $\Omega = 2$ and calculated modes with $\Omega = 3$ to 4 (see Chapter III). One needs to keep in mind that a higher Ω requires a higher value of the alternating field in the cavity which can become a limiting factor and in any case will require a substantial increase in the power needed by the radiofrequency generator. Thus, for $\Omega = 4$ the power loss in the cavity is 3-4 MW.

From what has thus far been said we can consider the creation of an effective single cavity microtron at 100 MeV entirely feasible.

Since the maximum energy is limited, further progress in acceleration of this type is connected with the increase of the current and the average power in the beam. The current in the beam depends first of all on the power of the high frequency generator. As the current grows, the collective effects of electron interactions become stronger, in particular their coupling with the oscillations in the cavity, and their coherent radiation during circular motion in the magnetic field (see Sections 3 and 6, Chapter VI). The last effect makes the limiting current about 1 A, where already the collective effects of the electron interactions become essential. This result, of course, requires experimental verification which can only be performed in the pulsed mode.

A large increase in beam power can be achieved by the use of a cw microtron or one operating in a long pulse mode (Chapter VIII). Such machines will greatly increase the possibilities in nuclear physics. Extremely attractive are cw microtrons with superconducting cavities.

We have had occasion to compare the microtron with other accelerators more than once in this book. It is timely to say a few more words about this. Existing electrostatic electron accelerators give a maximum energy of 5-8 MeV and a power comparable to that of pulsed microtrons. The only advantage of electrostatic accelerators is that their beam is more monoenergetic. However, in most cases, this is not very important since the monoenergeticity which the microtron is capable of is quite adequate.

At low energies the microtron has a decided advantage over the betatron. Although the technical complexity of the microtron is comparable to the betatron it provides a much more powerful beam (100-300 times larger) which is monoenergetic, well collimated, and can be easily extracted. The energy of this beam can be continuously varied within wide limits.

A more serious competitor of the microtron in the medium energy range (30-100 MeV) is the electron linac. The acceleration mechanism of the linac is similar to that of the microtron. The linac can be analyzed as a large number of cavities connected in series, while the microtron as a parallel connected system in which only one cavity (with a correspondingly lower characteristic resistance) accelerates all the electrons to the maximum energy.

For a given magnetic field in a microtron, acceleration is only possible if the power available at the cavity exceeds a certain threshold value (Section 1, Chapter VII). In a linac there is no such condition, however the power required by it (for the same electron energy) is noticeably larger than that required by the microtron cavity. A linac can provide pulsed acceleration of large currents because of its capability of storing electromagnetic energy. In a microtron, because of the necessity of accelerating particles on a fixed energy, this cannot be done. However, the accelerated beam in the microtron is much more monoenergetic than it is in the linac.

In comparison with the linac the striking advantage of the microtron is its simplicity. The microtron cavity, the accurate construction of which does not present any particular difficulty, is much simpler than the irised waveguide of the linac. In a microtron it is possible to use powerful self-exciting magnetrons which are unique in the reliability and simplicity of operation. The construction of the magnet for the microtron is simple and the energy losses in it are minimal.

We have already mentioned that the results presented in this book are based on research performed at the Institute for Physical Problems. The research would not have been possible without the support and consultation of P. L. Kapitza, for which the authors are deeply indebted.

We are indebted to L. A. Wainstein not only for valuable advice concerning this effort but also for the difficult task of editing this book, without which it would have hardly appeared.

The authors thank G. P. Prudkowsky and I. G. Krutikova who performed the calculations for particle motion in the initial stages of this research.

We recall with pleasure our association with V. P. Bikov in the construction of the first microtron.

We are obliged to S. I. Filimonov and A. I. Pavlov for solving many of the organizational problems.

From the very start of this effort the following people participated: A. A. Kolosov, S. V. Melekhin, D. N. Balashov, V. V. Ghuchkov, L. G. Ghukov, and B. G. Korolkov. Design of the microtrons was continuously performed by L. M. Zikin and in former years by V. N. Lvov. In the construction and the investigation of the machines we are deeply indebted to A. G. Nedelayev and B. S. Zakirov. In the later years the role contributed by our former students and present colleagues E. L. Kosarev, L. B. Lugansky, E. A. Lukianenko, V. M. Chernenko, and Yu. M. Tsipeniuk has become more important. We heartily thank these friends and other members of the Institute for Physical Problems without whom the ideas of the authors would not have materialized into the design and construction of the microtrons.

We would also like to acknowledge the cooperation of the microtron group of the Laboratory of Neutron Physics in Dubna, headed by I. M. Matora and R. V. Kharuzov.

The authors would finally like to express their thanks to Dr. E. Rowe who not only built the first successful modern microtron in the USA, but also helped in arranging for the translation of this book by Dr. I. N. Sviatoslavsky. We would also like to thank Mrs. Winterbottom for the preparation of the final version and last, but not least, Dr. John Blewett for his untiring effort in making the English translation available to the accelerator community, a community that owes so much to John's many sided activities in building and writing about accelerators.

APPENDIX I

DESIGNING THE MICROTRON

Engineering design calculations are usually done step by step with the aim of finding the best compromise, when several versions are usually calculated. We, however, will perform the calculations only once in order to demonstrate the basic logic of design.

Let us take as an example a pulsed microtron, operating at the 5 cm wavelength, which is to be used for nondestructive testing and thus has to be able to supply a dose of 1000 R/min at a distance of 1 m at an energy of 10 MeV.

The mode of acceleration and the value of Ω are selected according to Table 3.1. Because of the short wavelength, $\lambda = 5$ cm, the field strengths are considerable. For this reason we will select the first type of acceleration. Since in industrial radiography there is no need to vary the energy while the current in the beam should be as large as possible, we will select regime No. 6 (see Table 3.1) in a circular cavity with a large capture for $\Omega = 1.2$, $\varepsilon = 1.04$, $\ell = 1.07$ and $x_0 = 1.70$.

The cyclotron field H_0 for $\lambda = 5$ is equal to*

$$H_0 = \frac{10700}{\lambda \text{ (cm)}} = 2140 \text{ Oe} \tag{2.8}$$

and the constant magnetic field H for $\Omega = 1.2$ is

$$H = \Omega H_0 = 2560 \text{ Oe} .$$

Thus the amplitude of the electric field at the axis of the cavity is

$$E = \varepsilon H = 740 \text{ kV/cm} . \tag{2.7}$$

This value of E has to be considered allowable, although rather large.

The geometry of the cavity, its diameter, thickness, and the distance between the emitter and the axis are thus equal to

$$2a = 0.735 \lambda = 38 \text{ mm} , \tag{3.11}$$

---

*Equation numbers are taken from the text.

MICROTRON

$$L = \ell \frac{\lambda}{2\pi} = 8.5 \text{ mm} \quad , \quad (2.2)$$

$$X_0 = x_0 \frac{\lambda}{2\pi} = 8 \text{ mm} \quad . \quad (2.2)$$

The power of excitation from Table 3.1 is given for $\lambda = 10$ cm. For $\lambda = 5$ cm the power, according to Table 6.1 is

$$P_{r,\lambda} = P_{r,10} \sqrt{\frac{10}{\lambda}} = 35 \text{ kW} \quad ,$$

a value which can be made more accurate by the use of Eq. (6.9).

For a given value of bremsstrahlung the current needed in the beam can be evaluated from Fig. 7.33. For a total energy $U_N = 10.5$ MeV, the dose rate of 1000 R/min at a distance of 1 m can be had if the average current is 25 μA. Thus, for a duty ratio of 1000, the pulsed beam current has to be 25 mA at a power $P_N = 250$ kW. If, for purposes of discussion we assume the efficiency of the mode of acceleration $\eta_N = 50\%$, then the power needed at the cavity is

$$P = P_r + \frac{P_N}{\eta_N} = 850 \text{ kW} \quad , \quad (6.19)$$

and the coupling coefficient of the unloaded cavity to the waveguide is

$$\beta = \frac{P_N}{\eta_N P_r} + 1 = 2.4 \quad . \quad (6.14)$$

The Q factor of a cavity with a diameter 2a and thickness L is

$$Q = \frac{aL}{d(a + L)} = 7200 \quad , \quad (6.6)$$

where $d = 1.2 \sqrt{10/\lambda} \times 10^{-3}$ mm $= 0.85 \times 10^{-3}$ is the skin depth. Consequently, the loaded Q factor for a coupling coefficient of $\beta = 2.4$ will be

$$Q_H = \frac{Q}{3.4} = 2100 \quad . \quad (6.11)$$

The data given thus far is adequate for the selection of a generator and the design of the waveguide.

We will now consider the magnet, the field in which must be equal to $H = 2560$ Oe.

The number of orbits

$$N = \frac{U_N}{\Omega U_0} - 1 = 16 \quad . \quad (2.12)$$

Orbit separation is

$$\Delta D = \frac{\lambda}{\pi} = 16 \text{ mm} \qquad (1.12)$$

and the diameter of the last orbit is then

$$D = (N + 1) \frac{\lambda}{\pi} = 270 \text{ mm} \qquad (3.10)$$

We will assume that the distance between poles is

$$h = \lambda = 50 \text{ mm}$$

With such a separation between the magnet poles there will be plenty of room for the cavity and various pumping elements. However, more detailed design may change h.
The diameter of the magnet pole D_M may be taken as

$$D_M = D + h = 320 \text{ mm}$$

With a gap h = 5 cm the magnetomotive force of the coils must be

$$In = \frac{Hh}{0.4\pi} = 10,000 \text{ ampere-turns} \qquad (5.1)$$

if we ignore magnet saturation.
At a density i = 3 A/mm², the cross section of the copper conductor must be

$$S = \frac{In}{i} = 3330 \text{ mm}^2$$

If the coil consists of two sections (one section on each pole) with the height a = 20 mm, the width of the copper must be

$$b = \frac{S}{2a} = 84 \text{ mm}$$

For a radial packing coefficient of $\eta_r = 0.9$, the radial size of the coil b' is

$$b' = \frac{b}{\eta_r} = 93 \text{ mm}$$

The power needed to excite the coils at unit current density is proportional to its volume and is

$$W = \pi\rho(D + b')S \cdot j^2 = 780 \text{ W} \ ,$$

where $\rho = 1.8 \times 10^{-6}$ $\Omega \cdot$cm, the resistance of copper.
We can now select the power supply and supplement the coil data. Thus, for I = 30 A, the coil voltage will be

$$V = \frac{W}{I} = 26 \text{ V}$$

and the coil resistance will be R = V/I = 0.84 Ω. For the given parameter of the power supply the total number of turns n is equal to

$$n = \frac{In}{I} = 332 \quad .$$

Consequently, each section consists of 166 turns and can be wound with conductor of cross section

$$S = \frac{I}{i} = 10 \text{ mm}^2 \quad .$$

By using a ribbon 20 × 0.5 mm² the width of insulation must be 0.05 mm, due to the values of n_r and b' selected.
The coil data will determine the size of the aperture in the yoke; at this time one must select a place for cooling the coils and for attaching them.
The size of the yoke we will determine from the maximum induction B_{max} = 13,000 G, which we will assume for a magnet of Section AA (see diagram). This cross section must carry

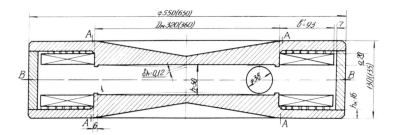

Magnet cross section.

the total flux. By equating the magnetic flux through the pole to the flux in Section AA we obtain

$$h_A = \frac{HD_M}{4B_{max}} = 16 \text{ mm} \quad .$$

The cross section of the backleg BB can be somewhat smaller. Demagnetization factors in these calculations have been ignored.
The profile of the poles near the edges are best determined by comparison with existing magnets (see Fig. 5.4); the dimensions of the Rose ring are about 6 × 1 mm² (see diagram). The sagitta depth δh can be estimated by the equation

$$\delta h = \frac{DB}{h_A \mu} h = 0.12 \text{ mm} \quad ,$$

where steel permeability is taken as μ = 2000. The gap tolerance is then

$$\Delta h \approx \left. \frac{h}{3N^2} \right| = 0.06 \text{ mm} , \qquad (4.57)$$

which determines the tolerance on the transverse field distortion.

However, the given values and the constants of the magnet

$$\frac{H}{I} = \frac{0.4\pi n}{h} = 83 \text{ Oe/A}$$

are best fixed after field mapping. We should mention that in magnets of this type, there is very little flux leakage, so that the measured constant H/I is usually within 1-2% of the calculated value. For this reason there is no need to take into account demagnetizing factors at the yoke.

By using this method we can obtain a magnet which will operate in the linear part of the magnetization curve for steel. The weight of such a magnet is somewhat more than the weight of an optimized magnet designed for this field.

A cross section of the accelerator is shown in the drawing. The weight of the magnet is 160 kg and the outer diameter is 550 mm.

While performing the calculations it is important that one is aware how the variation in the existing parameters will affect the characteristics of the machine. For this reason we have shown in the drawing values in brackets which give dimensions and other parameters for an energy of 13 MeV and 20 orbits (for the same Ω). At the same current and this energy level the radiation dose will double in value, the diameter of the poles will be 360 mm and the magnet diameter will be 600 mm. The whole machine will weight 200 kg. In order to get a better feeling for the variations in the accelerator characteristics it is also necessary to take into account the theoretical results obtained for the microtron (see Section 4, Chapter VI).

Further optimizing of the dimensions should be carried out with more of the necessary details of the accelerator construction already fixed. Vacuum chamber calculations, cooling of the cavity and the target, the steps for stabilizing the power supply, are carried out in the same way as for any other electrophysical installation. We will not investigate these problems, especially since their solution depends to a great extent on tradition, operating methods, and production technology.

The calculations for the accelerator illustrate a logical sequence from which can be seen the degree of the interaction between the parameters as well as the feedback of the obtained results on the given values. The accelerator builder should be well acquainted with these interactions, since in their understanding, to a large extent, lies the key to the success of the project.

APPENDIX II

MOTION CHARACTERISTICS AS EXPRESSED BY A SECOND ORDER MATRIX

On the outer orbits, where terms of the order proportional to $1/n$, $1/n^2$ ... (n = number of orbits) can be neglected, particle motion in the microtron on each revolution is determined by the matrix expression

$$\begin{pmatrix} z_{n+1} \\ p_{n+1} \end{pmatrix} = \begin{pmatrix} \alpha_{11} & \alpha_{12} \\ \alpha_{21} & \alpha_{22} \end{pmatrix} \cdot \begin{pmatrix} z_n \\ p_n \end{pmatrix} + \begin{pmatrix} \sigma_1 \\ \sigma_2 \end{pmatrix} \quad . \quad (II.1)$$

Here z_n and p_n are the coordinates and the momentum corresponding to a given motion, (α_{ik} is a matrix with constant elements, and σ_1 and σ_2 the total increment in the coordinate and the momentum on each revolution caused by external forces which do not depend on the coordinate or the momentum σ_1 and σ_2 take into account the action of the fixed magnetic field, the effect of cavity misalignment, etc.).

For $\alpha_{12} \neq 0$, expression (II.1) can be replaced by the equivalent difference equations

$$z_{n+1} = Sz_n - Dz_{n-1} + (1 - \alpha_{22})\sigma_1 + \alpha_{12}\sigma_2$$

$$p_n = (z_{n+1} - \alpha_{11}z_n - \sigma_1)/\alpha_{12} \quad , \quad (II.2)$$

where $S = \alpha_{11} + \alpha_{22}$ is the matrix trace, and $D = \alpha_{11}\alpha_{22} - \alpha_{12}\alpha_{21} = \text{Det}(\alpha_{ik})$ is the matrix determinant. In order for z and p to exchange places in the formulas (II.2), all the values α_{ik} and σ_i should exchange indices $1 \rightleftarrows 2$.

It is important to note that the matrices encountered in this book satisfy the relationship

$$D = 1 \quad , \quad (II.3)$$

which is the result of the fact that the equation of motion after linearization has the form

$$\frac{dz}{d\phi} = f_1(\phi)p \quad ,$$

$$\frac{dp}{d\phi} = f_2(\phi)z + f_3(\phi) \quad .$$

Let us write the solution a

$$\begin{pmatrix} z \\ p \end{pmatrix} = \begin{pmatrix} t_{11} & t_{12} \\ t_{21} & t_{22} \end{pmatrix} \begin{pmatrix} z_0 \\ p_0 \end{pmatrix} + \begin{pmatrix} \hat{z} \\ \hat{p} \end{pmatrix} ,$$

where z_0 and p_0 are the initial values of position and momentum (for $\phi = 0$), z and p the same quantities at an arbitrary moment ϕ, $\hat{z}$ and $\hat{p}$ are the special solutions of the nonhomogeneous equation which satisfy zero initial conditions, and $t_{ik} = t_{ik}(\phi)$ is the fundamental system of solutions for the homogeneous equation obtained for $f_3 = 0$. Accordingly, the functions t_{ik} are satisfied by the equation

$$\frac{dt_{11}}{d\phi} = f_1 t_{21} , \qquad \frac{dt_{12}}{d\phi} = f_1 t_{22} ,$$

$$\frac{dt_{21}}{d\phi} = f_2 t_{11} , \qquad \frac{dt_{22}}{d\phi} = f_2 t_{12} ,$$

and the initial conditions

$$t_{11} = t_{22} = 1 \quad \text{and} \quad t_{22} = t_{21} = 0 \quad \text{for } \phi = 0 .$$

From this

$$\frac{d}{d\phi} (t_{11} t_{22} - t_{12} t_{21}) = 0$$

and according to the initial conditions

$$\text{Det}(t_{ik}) = 1 .$$

For the value ϕ_1 which corresponds to a complete revolution, t_{ik} becomes α_{ik} and we obtain Eq. (II.3). This relationship insures the conservation of phase volume which is manifested, in the general case, by the motion of charged particles in a given electromagnetic field.[48] Hereafter we will consider $D = 1$.

Let us investigate the homogeneous differential equation ($\sigma_1 = 0$). The first equation (II.2) will be written as

$$z_{n+1} = S z_n - z_{n-1} .$$

It has the solution

$$z_n = \lambda^n , \qquad \lambda_{1,2} = \frac{S}{2} \pm \sqrt{\frac{S^2}{4} - 1} . \qquad (II.4)$$

From (II.4) it is evident that for

$$-2 < S < 2 \qquad (II.5)$$

stable oscillations occur at a frequency ν

$$z_n = C_1 \cos \nu n + C_2 \sin \nu n ,$$

MICROTRON   197

where

$$\cos \nu = S/2 \quad , \quad 0 < \nu < \pi \quad . \tag{II.6}$$

The values S, determined by the inequality (II.5), correspond to stable focusing in accelerators (see Fig. 2.5). Values S > 2 lead to defocusing; one of the solutions increases exponentially as n gets larger (see Fig. 2.3). For S < -2 particle motion is also unstable; the value z_n changes sign during the transition of n to n + 1, at which time the amplitude of the oscillations increases (see Fig. 2.4). From (II.4) it is evident that $\lambda_{1,2}$ for $S = -|S|$ differs in sign from the value $\lambda_{2,1}$ for $S = |S|$. Thus, in effect, the growth in the amplitude of the oscillations for S < -2 goes according to the same law as the monotonic increase during defocusing.

Let us investigate the properties of stable motion. From the condition

$$\alpha_{11}\alpha_{22} - \alpha_{12}\alpha_{21} = 1 \quad , \quad -2 < \alpha_{11} + \alpha_{22} < 2$$

it follows that $-\alpha_{12}\alpha_{21} > 0$. Introducing

$$\operatorname{sgn} \alpha_{21} = \frac{\alpha_{21}}{|\alpha_{21}|} \quad (\text{sign } \alpha_{21})$$

we determine the values ρ and ζ according to

$$\rho = \operatorname{sgn} \alpha_{21} \sqrt{-\frac{\alpha_{12}}{\alpha_{21}}} \quad ,$$

$$\cos \zeta = \frac{\alpha_{11} - \alpha_{22}}{2\sqrt{-\alpha_{12}\alpha_{21}}} \quad , \quad \sin \zeta = \frac{\sin \nu}{\sqrt{-\alpha_{12}\alpha_{21}}} \quad ,$$

$$0 < \zeta < \pi \quad .$$

Then from (II.6) and the second part of (II.2) it follows

$$p_n = C \cos (\nu n + \chi) \quad ,$$

$$z_n = \rho C \cos (\nu n + \chi + \zeta) \quad . \tag{II.7}$$

In the z_n, p_n plane the locus of points depicts an ellipse, the equation of which can be written in several ways,

$$z_n^2 - 2\rho (\cos \zeta) z_n p_n + \rho^2 p_n^2 = \rho^2 C^2 \sin^2 \zeta \quad ,$$

$$\alpha_{11} z_n^2 - (\alpha_{11} - \alpha_{22}) z_n p_n - \alpha_{12} p_n^2 = C^2 \frac{\sin^2 \nu}{\alpha_{21}} \quad . \tag{II.8}$$

The ellipse in the z,p plane goes counterclockwise for $\rho > 0$ and clockwise for $\rho < 0$.

From (II.8) it is evident that for the same initial conditions the amplitude of oscillations is larger as the system approaches the stability boundary. If $|S| \to 2$, then $\sin \zeta \to 0$

and $C \to \infty$. The amplitude of oscillations generally speaking is smallest in the center of the stability band, i.e., when $S \approx 0$.

Equations (II.4) to (II.8) are obtained by us from the difference equations (II.2) and in their derivation it was assumed that $\alpha_{12} \neq 0$. However, for $\alpha_{12} = 0$ from the condition (II.3) it follows that $\alpha_{11} = \alpha_{22} = \pm 1$, i.e., $|S| = 2$, whereas for Eqs. (II.4) to (II.8) this case does not apply.

Let the value α_{12} as previously be different from zero and $S = 2$. Equation (II.4) only gives us the solution $z_n =$ const, while the other independent linear solution, as can be easily seen, is proportional to n. The general solution of the system (II.2) has the form

$$z_n = C_1 + C_2 n ,$$

$$p_n = \frac{1 - \alpha_{11}}{\alpha_{12}} (C_1 + C_2 n) + \frac{C_2}{\alpha_{12}} . \qquad (II.9)$$

From condition (II.3) it follows that for $S = 2$ and $\alpha_{12} \neq 0$, then $\alpha_{11} = 1$ when, and only when, $\alpha_{21} = 0$. Thus, if all the elements of matrix (α_{ik}) are different from zero, then for $S = 2$ both values (z_n and p_n) increase linearly according to Eq. (II.9). If, however, $\alpha_{21} = 0$, then the homogeneous Eq. (II.1) takes on the form

$$\begin{pmatrix} z_{n+1} \\ p_{n+1} \end{pmatrix} = \begin{pmatrix} 1 & \alpha_{12} \\ 0 & 1 \end{pmatrix} \begin{pmatrix} z_n \\ p_n \end{pmatrix}$$

and has the solution

$$z_n = C_1 + C_2 n ,$$

$$p_n = C_2/\alpha_{12} = \text{const} . \qquad (II.10)$$

For $\alpha_{12} = 0$ and $\alpha_{21} \neq 0$ we have the unit matrix and the solution is obvious; the values z_n and p_n are constant. For completeness we should note that for $S = -2$ then (II.9) becomes

$$z_n = (-1)^n (C_1 + C_2 n) ,$$

$$p_n = (-1)^{n+1} \left[\frac{1 + \alpha_{11}}{\alpha_{12}} (C_1 + C_2 n) + \frac{C_2}{\alpha_{12}} \right] . \qquad (II.11)$$

Equations (II.10) also change correspondingly, since the values z_n and p_n have to be multiplied by $(-1)^n$ and $(-1)^{n+1}$.

Let us now examine the action of external forces described by the terms σ_i in Eqs. (II.1) and (II.2). The difference equation for z_n will become

$$z_{n+1} = S z_n - z_{n-1} + F ,$$

$$F = (1 - \alpha_{22}) \sigma_1 + \alpha_{12} \sigma_2 . \qquad (II.12)$$

For S ≠ 2 the partial solution of Eq. (II.12) has the form

$$z_n = \frac{F}{2 - S} = \text{const} \tag{II.13}$$

and represents a shift in the center of oscillation under the action of external forces. If S = 2, then solution (II.13) is replaced by

$$z_n = \frac{F}{2} n^2 \quad . \tag{II.14}$$

In calculating vertical focusing, one encounters the difference equation

$$z_{n+1} = \left(2 - \frac{\delta}{n}\right) z_n - z_{n-1} + F \quad , \tag{II.15}$$

in which $S_n = 2 - \delta/n \to 2$ for $n \to \infty$. Partial solution of this equation has the form

$$z_n = \frac{F}{\delta} n \quad . \tag{II.16}$$

BIBLIOGRAPHY

SECTION I

1. S. P. Kapitza, "A method for bunching fast electrons," *JTP*, 1959, 29, 729.
2. S. P. Kapitza, V. P. Bikov, V. N. Melekhin, "Microtron for high current," *JETP*, 1960, 39, 997.
3. S. P. Kapitza, V. P. Bikov, V. N. Melekhin, "Efficient microtrons for high intensity," *JETP*, 1961, 41, 368.
4. S. P. Kapitza, V. N. Melekhin, I. G. Krutikova, G. P. Prudkovsky, "Electron orbits in microtrons," *JETP*, 1961, 41, 376.
5. V. N. Melekhin, "Vertical focusing in microtrons," *JETP*, 1962, 42, 622.
6. S. P. Kapitza, L. A. Wainstein, "Radiation damping of electron motion in microtrons," *JETP*, 1962, 42, 821.
7. V. P. Bikov, "Investigation of electron bunching in microtrons," *JETP*, 1961, 40, 1658.
8. V. P. Bikov, "Electron bunching in microtrons," *JETP*, 1963, 44, 1425.
9. V. P. Bikov, "Effects of magnetic field variations on the motion of beams in microtrons," *JTP*, 1963, 33, 337.
10. L. M. Zykin, S. P. Kapitza, V. N. Melekhin, A. G. Nedelyaev, "Microtrons for high intensity," *Proceedings International Conference on Accelerators, Dubna*, 1963, p. 1049.
11. V. N. Melekhin, "Dynamics of electrons in microtrons," *Proceedings International Conference on Accelerators, Dubna*, 1963, p. 1053.
12. S. P. Kapitza, "The Microtron," Thesis, Joint Institute for Nuclear Research, 1961.
13. V. P. Bikov, "Bunching of electrons in the microtron," Thesis, Institute of Physical Problems, 1961.
14. V. N. Melekhin, "Dynamics of electrons in microtrons," Thesis, Institute of Physical Problems, 1961.
15. S. P. Kapitza, "New sources of fast electrons," *Proc. USSR Acad. Sci.*, 1961, No. 10, 65.
16. S. P. Kapitza, "Microtrons and their applications," *Atomnaya Energia*, 1965, 18, 203.
17. S. P. Kapitza, "Microtron applications," in the collection: *Electronics at High Energies*, Vol. 4, 1965, p. 178.
18. S. P. Kapitza, "The cw microtron," *V International Conference on High Energy Accelerators, Frascati*, 1965, p. 672.
19. S. P. Kapitza, E. L. Kosarev, E. A. Lukyanenko, V. N. Melekhin, A. G. Nedelyaev, V. M. Chernenko, L. M. Zykin, "The 30 MeV high current microtron," *V International*

Conference on High Energy Accelerators, Frascati, 1965, p. 671.
20. S. P. Kapitza, "Modern developments of the microtron," V International Conference on High Energy Accelerators, Frascati, 1965, p. 665.
21. V. N. Melekhin, "Microtron efficiency," in the collection: Electronics at High Energies, Vol. 5, 1968, p. 228.
22. L. B. Lugansky, V. N. Melekhin, "Numerical analysis of the acceleration process in the microtron," in the collection: Electronics at High Energies, Vol. 5, 1968, p. 238.
23. V. N. Melekhin, E. A. Lukyanenko, "Electron dynamics in microtrons with nonuniform magnetic fields," in the collection: Electronics at High Energies, Vol. 5, 1968, p. 257.
24. E. A. Lukyanenko, V. N. Melekhin, V. M. Chernenko, "Measurement of the nonuniformity of the magnetic field in a 30-orbit microtron," in the collection: Electronics at High Energies, Vol. 5, 1968, p. 275.
25. E. L. Kosarev, "Phase oscillations in a microtron of high intensity," in the collection: Electronics at High Energies, Vol. 5, 1968, p. 283.
26. E. L. Kosarev, "Measurement of the accelerator cavity in a microtron," in the collection: Electronics at High Energies, Vol. 5, 1968, p. 306.
27. S. P. Kapitza, V. N. Melekhin, B. S. Zakirov, L. M. Zykin, E. A. Lukyanenko, Yu. M. Tsipeniuk, "Construction and test of a 17-orbit microtron," Instruments and Experimental Techniques (PTE), 1969, No. 1, 13.
28. V. D. Ananev, S. P. Kapitza, I. M. Matora, V. N. Melekhin, L. A. Merkulov, R. V. Kharuyzov, "A 30-MeV microtron injector for I. B. Ra, Atomnaya Energia, 1966, 20, 106.
29. S. P. Kapitza, Yu. M. Tsipeniuk,"Measurement of high frequency fields in resonators," in the collection: Electronics at High Energies, Vol. 2, 1963, p. 133.
30. P. L. Kapitza, S. J. Philimonov, S. P. Kapitza, "Nigotron," in the collection: Electronics at High Energies, Vol. 6, 1969, p. 7.
31. G. P. Prudkovsky, "Orbit tracer," in the collection: Electronics at High Energies, Vol. 3, 1964, p. 70.
32. F. S. Rusin, G. D. Bogomolov, "Generation of electromagnetic oscillations in open resonators," Letter in JETP, 1966, 4, No. 6, 236.
33. S. P. Kapitza, "Reasonable system of units in classical electrodynamics and electronics," Progress of Physical Science (UPN), 1966, 88, 191.
34. A. S. Soldatov, G. N. Smirenkin, S. P. Kapitza, Yu. M. Tsipeniuk,"Quadrupole fission of U^{238}," Phys. Letters, 1965, 14, 217.
35. I. B. Bocharova, V. G. Zolotuchin, S. P. Kapitza, G. N. Smirenkin, A. S. Soldatov, Yu. M. Tsipeniuk,"Angular distribution of fragments at the threshold for photofission of U^{238}," JETP, 1965, 49, 476.
36. S. P. Kapitza, V. I. Novgorodtseva, V. P. Pchelin, G. N. Smirenkin, Yu. M. Tsipeniuk,M. P. Shubko, "On symmetrical fission of U^{238}, Letter in JETP, 1967, 6, 495.

37. S. P. Kapitza, "Laboratory electromagnet," *Instruments and Experimental Techniques*, 1958, 2, 97.
38. Yu. M. Tsipeniuk,"Investigation and application of electron and γ-ray beams from microtrons," Thesis, Institute of Physical Problems, 1968.
39. F. S. Rusin, "Linear theory of the orotron," in the collection: *Electronics at High Energies*, Vol. 5, 1968, p. 9.
40. L. B. Lugansky, "New region of phase oscillations in microtrons," in the collection: *Electronics at High Energies*, Vol. 6, 1969, p. 130.

SECTION II

41. V. I. Veksler, *Proc. USSR Acad. Sci.*, 1944, 43, 346; *J. Phys. USSR*, 1945, 9, 153.
42. W. J. Henderson, H. Le Caine, R. Montalbetti, *Nature*, 1948, 162, 699.
43. P. A. Readhead, H. Le Caine, W. J. Henderson, *Canad. J. Res.*, 1950, A28, 73.
44. A. P. Greenburg, *Progress of Physical Sciences (UPN)*, 1961, 75, 421.
45. A. Roberts, *Ann. Phys.*, 1958, 4, 115.
46. A. A. Kolomensky, Thesis, Lebedev Institute (PIAN), 1950.
47. A. A. Kolomensky, *JTP*, 1960, 30, 1347.
48. P. A. Sturrock, *Static and Dynamic Electron Optics*, Moscow Foreign Literature Press, 1958.
49. C. Henderson, F. F. Heyman, R. E. Jennings, *Proc. Phys. Soc.*, 1953, B66, 654; 1953, B66, 41.
50. H. F. Kaiser, *Rev. Sci. Instrum.*, 1954, 25, 1025; *J. Franklin Inst.*, 1955, 259, 25.
51. E. Kiski-Koszo, *Acta phys. Acad. scient hung.*, 1955, 4 377.
52. A. Carrelli, F. Porecca, *Nuevo Cimento, Suppl.*, 1957, 6, 729.
53. H. Reich, *Z. angew. Phys.*, 1960, 12, 481.
54. A. Paulin, *Nucl. Instrum. and Methods*, 1959, 5, 107; 1960, 9, 113; 1962, 15, 306.
55. C. Schmeltzer, *Z. Naturforschung*, 1952, 7a, 808.
56. L. A. Wainstein, *Electromagnetic Waves*, published by "Soviet Radio," 1957.
57. D. K. Aitken, *Proc. Phys. Soc.*, 1957, A70, 550.
58. H. Reich, K. Löns, *Nucl. Instrum. and Methods*, 1964, 31, 221.
59. O. Wernholm, *Arkiv. fys.*, 1964, 26, 527.
60. D. K. Aitken, F. F. Keymann, R. E. Jennings, P. I. P. Kalmus, *Proc. Phys. Soc.*, 1961, 77, 769.
61. J. S. Bell, *Proc. Phys. Soc.*, 1953, B66, 802.
62. E. M. Moroz, *Translations of Lebedev Institute (PIAN)*, 1966, 13; *Proc. USSR Acad. Sci.*, 1956, 108, 436; 1957, 115, 78; *Atomnaya Energia*, 1958, 4, No. 1, 238.
63. E. Brannen, H. Froelich, *J. Appl. Phys.*, 1961, 32, 1179; *IEEE Trans.*, MTT-11, 1963, 5, 288.
64. J. Rubi, "The magnet," 1967, No. 8.
65. Y. R. Davies, R. E. Jennings, F. Porecca, R. E. Rand, *Nuovo Cimento, Suppl.*, 1960, 17, No. 2, 202.

66. K. A. Belovintsev, A. Ya. Belyak, S. B. Vorontsov, P. A. Cherenkov, *Proc. International Conf. on Accelerators*, Dubna, 1963, p. 1056.
67. K. A. Belovintsev, P. A. Cherenkov, *ibid.*, p. 1061.
68. K. A. Belovintsev, F. P. Denisov, *Atomnaya Energia*, 1964, 16, 353.
69. B. Z. Kanter, Yu. P. Yushkov, *Instruments and Experimental Techniques (PTE)*, 1964, No. 4, 28.
70. V. A. Povitski, A. I. Tsarapaev, *JTP*, 1964, 34, 1462.
71. O. Wernholm, private communication.
72. G. A. Kudintseva, *Electronics*, Series 1, 1960, No. 4, 81.
73. I. M. Matora, P. B. Kharyuzov, *Instruments and Experimental Techniques (PTE)*, 1967, 3, 28.
74. V. P. Stepanchuk, *Transactions of Saratov Students, University Physics Proceedings*, 1965, pp. 65-74, 187.
75. V. P. Stepanchuk, Thesis, Saratov, 1966.
76. V. P. Stepanchuk, "Electron accelerators," *Proc. of the Fifth Interuniversity Conf., Tomsk*, 1964, Moscow Atomizdat, 1966.
77. K. A. Belovintsev, A. Ya. Belyak, A. M. Gromov, E. M. Moroz, P. A. Cherenkov, *Atomnaya Energia*, 1963, 14, 359.
78. O. S. Milovanov, Thesis, Moscow Engineering Physics Institute, 1962.
79. O. A. Valdner, "Electron linear accelerator" (basic calculations on electron linear accelerators at low energies), Moscow Atomizdat, 1966.
80. W. Sells, H. Froelich, E. Brannen, *J. Appl. Phys.*, 1965, 36, 3264.
81. W. Sells, F. C. Choo, E. Brannen, H. R. Froelich, *Proc. IEEE*, 1966, 54, 552.
82. B. Z. Kanter, *Instruments and Experimental Techniques (PTE)*, 1965, No. 3, 34-38.
83. B. H. Wiik, H. A. Schwettman, P. B. Wilson, *V. International Conference on High Energy Accelerators, Frascati*, 1965.
84. A. Campolattaro, F. Porreca, *Rev. Sci. Instrum.*, 1967, 38, 1322.
85. P. P. Wintersteiner, D. S. Edmonds, Jr., *IEEE Trans. Nucl. Sci.*, 1967, NS-14, No. 3, 749.
86. H. R. Froelich, E. Brannen, *IEEE Trans. Nucl. Sci.*, 1967, NS-14, No. 3, 756.
87. C. S. Robinson, D. Jamnik, A. O. Hanson, *IEEE Trans. Nucl. Sci.*, 1967, NS-14, No. 3, 624.
88. F. Amman, M. Bernardini, et al., *V. International Conference on High Energy Accelerators, Frascati*, 1965, p. 703.
89. V. D. Ananev, P. S. Antsupor, I. M. Matora, L. A. Merkulov, R. V. Kharyuzov, Preprint, Joint Institute for Nuclear Research, 9-3283, Dubna, 1967.
90. K. A. Belovintsev, Thesis, Institute for Physical Problems, 1964.
91. L. W. Alvarez, R. Cornog, *Phys. Rev.*, 1939, 56, 379.
92. A. A. Kolomensky, Letter in *JETP*, 1967, 5, Part 6, 204.
93. V. Bizzarri, A. Vignati, *Nuovo Cimento*, 1970, 68A, 513.
94. L. B. Luganski, V. N. Melekhin, *JTP*, 1973, 43, 1954.
95. F. V. Rodionov, V. P. Stepanchuk, *JTP*, 1971, 41, 999.
96. V. N. Melekhin, *JETP*, 1971, 61, 1319.

97. V. N. Melekhin, L. B. Luganski, *JTP*, 1970, **40**, 2465.
98. V. I. Polyakov, F. V. Rodionov, V. P. Stepanchuk, *JTP*, 1971, **41**, 1667.
99. V. N. Melekhin, *JETP*, 1975, **68**, 1601.
100. E. M. Rowe, F. E. Mills, *Particle Accelerators*, **4**, 211 (1973).